Powering the Future: Blueprint for a Sustainable Electricity Industry

CHRISTOPHER FLAVIN AND

NICHOLAS LENSSEN

David Malin Roodman, *Staff Researcher*

Carole Douglis, *Editor*

WORLDWATCH PAPER 119
June 1994

THE WORLDWATCH INSTITUTE is an independent, nonprofit environmental research organization based in Washington, D.C. Its mission is to foster a sustainable society—in which human needs are met in ways that do not threaten the health of the natural environment or future generations. To this end, the Institute conducts interdisciplinary research on emerging global issues, the results of which are published and disseminated to decisionmakers and the media.

FINANCIAL SUPPORT is provided by the Geraldine R. Dodge Foundation, Energy Foundation, W. Alton Jones Foundation, John D. and Catherine T. MacArthur Foundation, Andrew W. Mellon Foundation, Joyce Mertz-Gilmore Foundation, Edward John Noble Foundation, Pew Charitable Trusts, Lynn R. and Karl E. Prickett Fund, Rockefeller Brothers Fund, Turner Foundation, Frank Weeden Foundation, Wallace Genetic Foundation, and Peter Buckley.

PUBLICATIONS of the Institute include the annual *State of the World*, which is now published in 27 languages; *Vital Signs*, an annual compendium of the global trends—environmental, economic, and social—that are shaping our future; the *Environmental Alert* book series; and *World Watch* magazine, as well as the *Worldwatch Papers*. For more information on Worldwatch publications, write: Worldwatch Institute, 1776 Massachusetts Ave., N.W., Washington, DC 20036; or FAX (202) 296-7635.

THE WORLDWATCH PAPERS provide in-depth, quantitative and qualitative analysis of the major issues affecting prospects for a sustainable society. The Papers are authored by members of the Worldwatch Institute research staff and reviewed by experts in the field. Published in five languages, they have been used as a concise and authoritative reference by governments, nongovernmental organizations and educational institutions worldwide. For a partial list of available Papers, see page 75.

DATA from all graphs and tables contained in this book, as well as from those in all other Worldwatch publications of the past year, are available on diskette for use with IBM-compatible computers. This includes data from the *State of the World* series of books, *Vital Signs* series of books, Worldwatch Papers, *World Watch* magazine, and the *Environmental Alert* series of books. The data are formatted for use with spreadsheet software compatible with Lotus 1-2-3, including Quattro Pro, Excel, SuperCalc, and many others. Both 3 1/2" and 5 1/4" diskettes are supplied. To order, send check or money order for $89, or credit card number and expiration date (Visa and MasterCard only), to Worldwatch Institute, 1776 Massachusetts Ave., NW, Washington, DC 20036. Tel: 202-452-1999; Fax: 202-296-7365; E-mail: Worldwatch@igc.apc.org.

Library of Congress Catalog Number 94-060382

ISBN 1-878071-20-3

Printed on 100-percent non-chlorine bleached, partially recycled paper.

Table of Contents

Tables and Figures

ACKNOWLEDGMENTS: We would like to thank Gerald Braun, Ralph Cavanagh, Armond Cohen, S. David Freeman, Jan Hamrin, Richard F. Hirsh, Eric Hirst, David Lapp, Reinhard Loske, David Moskovitz, David Penn, Carl Weinberg, and Stephen Wiel for their comments on earlier drafts of this document, and Michael Scholand for his research assistance. We also would like to thank Denise Byers Thomma, our in-house production editor, for guiding this project from raw manuscript to final printing. Finally, we are grateful to David Malin Roodman for his help in gathering information, running database and spreadsheet programs, reviewing numerous drafts, and providing invaluable insights on an extraordinarily complex subject.

CHRISTOPHER FLAVIN is vice president for research at the Worldwatch Institute. **NICHOLAS LENSSEN** is a senior researcher at the Worldwatch Institute. Flavin and Lenssen are the authors of *Power Surge: Guide to the Coming Energy Revolution,* published by W.W. Norton & Company, October 1994.

Introduction

Long known for its vast scale and fierce resistance to change, the electric power industry is poised for a sweeping transformation in the years ahead. Although driven by many of the same technological and economic forces that are propelling the telecommunications revolution, the electric power industry has received only a fraction as much attention until now. Yet the electric industry is larger than the telecommunications business, and has a far greater impact on the health of the global environment.

For two decades, a rising tide of new technologies, environmental concerns, and political developments has chipped away at the electric power business, altering its practices but not its basic structure. Today, that process of incremental change is giving way to a more fundamental overhaul, with profound implications for billions of people, from the price they pay to light their homes to the quality of the air they breathe.

Although electricity provides only 13 percent of the world's end-use energy (17 percent in industrial countries), it is a form of energy that is indispensable to today's economies and lifestyles—from electronic communications to the production of aluminum. Generating and distributing electricity has become one of the world's largest businesses, with annual revenues estimated at more than $800 billion—roughly twice the size of the world auto industry. The multibillion dollar power plants that generate much of the world's electricity number among the world's largest machines, and the networks of power lines extending from those plants are like a vast web of arteries, providing vital services to millions of consumers.[1]

Along with electricity, the power industry generates some of the world's most serious environmental problems. The industry is the leading consumer of fossil fuels, particularly coal, the dirtiest of those fuels. Consequently, power generation accounts for nearly one-third of global emissions of carbon dioxide, the principle greenhouse gas, and produces nearly two-thirds of the sulfur dioxide that is a major culprit in both local and long-distance air pollution. In recent decades, the power industry has been implicated in a series of additional environmental problems, ranging from emission of heavy metals to toxic ash, land degradation from strip mining, and the buildup of radioactive waste. It is hardly surprising that in the two years since the Earth Summit in Rio de 26, the power industry has loomed large in the efforts of many countries to develop more sustainable energy strategies.[2]

Beyond demands for a cleaner environment, the power industry faces severe economic challenges. Though organized in a variety of ways in different countries, the power business is generally dominated by vertically integrated monopolies: In most cases, the same public or private enterprise owns the power plants, transmission lines, and local distribution wires. Over the years, the industry has built increasingly large, centralized power plants—most of them fueled by coal, hydropower, or nuclear energy. Although these technologies contributed to a steady decline in electricity prices for several decades, rising costs and mounting public opposition caused serious damage to many power companies during the seventies and eighties.

Even today, governments are deeply involved in the power systems of all countries—either owning or heavily regulating most of the companies involved. Consequently, it is governments that have underwritten the current structure of the industry, and that often subsidize electricity production and use. Some nations, for example, encourage excessive, uneconomic consumption of electricity; others compel or encourage utilities to burn domestically produced coal instead of switching to cleaner fuels.

To address all these problems simultaneously, and build the foundation of an environmentally sustainable power system, the very structure of that industry will need to be revised—a process that has already begun in several nations. Independent, unregulated companies are now building many of the new power plants in some countries, generally delivering electricity from facilities that are less expensive and less environmentally damaging than most utility plants. At the same time, many utilities are devoting a rising share of their investment capital to improving the efficiency of their customers' appliances, buildings, and factories—spending an estimated $2.8 billion in 1993 in the United States alone. This is transforming some utilities into service-focused companies that meet the needs of consumers with whatever mix of investments will best do the job. These innovations are driven in part by economic and political forces—including requirements in many U.S. states that utilities adopt integrated resource plans that consider the full range of options for meeting power needs.[3]

The industry is the leading consumer of fossil fuels, particularly coal, the dirtiest of those fuels.

The other major force at work is the advent of a new generation of technologies, including gas turbines, wind turbines, fuel cells, and solar generators. Each of these is a relatively small-scale, potentially mass-produced means of generating power, with the capacity to create a less expensive and more decentralized electricity system. According to some engineers, it will soon be common to generate power within individual buildings, reducing energy conversion and transmission losses, and increasing the overall reliability of the entire power system. At the same time, the advent of new electronic control technologies will allow thousands of individual generators and consumers to be integrated into a single "smart" system that balances supply and demand by automatically turning individual generating and consuming devices on and off.

Even as the new technologies develop and the power industry continues to experiment with other means of producing and conserving electricity, corporate, institutional, and regulatory changes are underway. Already, some governments have begun to open the power generation market to competition.

Others have sold government-owned systems to private investors or have broken up their existing power monopolies. A few have begun to consider a system called "retail wheeling" in which individual customers purchase power directly from generating companies.

Although most of the changes now underway in the power industry have had positive effects—both economically and environmentally—the forces now unleashed could move in several directions, some of which might do more harm than good. Of particular concern are proposals to create undifferentiated commodity markets in electricity, providing little incentive for utilities to improve end-use efficiency or shift to less environmentally damaging sources of electricity. Most proposals for retail wheeling, for example, would undermine demand-side management, which, in many countries, has turned out to be the least expensive way of meeting electricity needs.

In order to effectively harness the power of the market—at all levels of the electricity industry—a two-tiered reform is needed. At the wholesale level, an open, competitive market for power makes sense, with a range of companies competing for electricity contracts, utilizing an open-access transmission system. However, mechanisms are needed to ensure that costs associated with environmental damage are included in that market.

Even more importantly, the power distribution business should be clearly separated from supply, and operated under a different model. The key to maximizing benefits at the distribution level—which in most cases remains a natural monopoly—is to create a competitive market in electricity services, with the local distribution company serving as financier and facilitator for companies that wish to offer customers more efficient lights or refrigerators, or even a rooftop solar generator. These distribution utilities would be governed by a new system of regulatory incentives that rewards them not for the capital they spend, but for their success in providing low-cost services to customers. They would also be responsible for comprehensive long-term planning.

Given the diversity of the power industry and the complexity of the issues now raised, a long period of experimentation seems likely, as various governments pursue individual approaches to electric utility reform, and as different classes of power customers compete to get the best deal possible. The years ahead are likely to be fractious and confusing, but, in the end, a far more efficient and less environmentally destructive power industry is likely to emerge.

End of the Line

When Thomas Edison started the world's first electric power company in New York in 1880, it looked like a typical, underfinanced start-up venture, not much different from hundreds of other small businesses. In a Wall Street warehouse, Edison connected a coal-fired boiler to a steam engine and dynamo, then linked the plant by underground wire to a block of nearby office buildings. When the switches were flipped at the Pearl Street Station on September 6, 1882, 158 light bulbs (also designed by Edison) flashed on, and the Edison Electric Illuminating Company made converts of its carefully chosen first customers—J.P. Morgan and the *New York Times*.[4]

Some of the industry's early leaders envisioned a system of stand-alone, mass-produced generators, but they were opposed by a competing vision: central plants selling power to hundreds of customers.

Electric power was an immediate hit—turning its pioneer into a millionaire, and changing the lives of millions of people. Electric lighting brought a new nighttime brightness to cities around the world and soon pushed competing "gas light" companies into oblivion. On the factory floor, thousands of small electric motors rapidly replaced steam engines and elaborate mechanical drive systems. Electricity came to be seen as an essential tool of development—

and a symbol of modernity. Although electric power cost far more than other forms of energy, its ability to run a new generation of industrial equipment and household appliances left it with no real competitors.

Edison viewed electricity as a dynamic, competitive service business, initially even selling lighting to his customers by the bulb. Without government regulation or other controls, electric power companies quickly proliferated, offering both direct and alternating current at various voltages, and often running competing electric lines down opposite sides of the same street. Some of the industry's early leaders envisioned a system of stand-alone, mass-produced generators, but they were opposed by a competing vision: central plants selling power to hundreds of customers. The latter eventually prevailed, driven in part by the advent of alternating current and the transformer, which made it possible to raise the voltage of electric current and thereby transmit it over long distances. In addition, the rotary steam turbine, developed by Charles Parsons in the 1880s, generated power more efficiently and economically when built on a large scale, encouraging the trend to bigger plants. Water-driven turbines used in hydropower projects provided the other major source of electricity.[5]

Four dozen power companies served Chicago alone at the turn of the century. But in just a few decades, single utilities were serving whole cities, then entire regions. The industry's growth transformed it, for with monopoly control came an implied public responsibility. Beginning in 1907, state regulatory commissions formed in the United States to determine a fair price for electricity, and to provide financial stability for utilities.[6]

Similar consolidations took place around the world. Following World War II, the United Kingdom still had nearly 600 electricity companies, roughly one-third of them privately owned. Within a decade, the country possessed just one state-owned generating company—the Central Electricity Generating Board—which supplied bulk power to twelve regional distributors. Elsewhere, governments took over many electric utilities, leading eventually to national utilities in many countries, including France, Indonesia, and South Africa. Often, the resulting monopoly structures bore closer resemblance to the economic visions of Lenin than to the vigorous capitalism of Edison's late-nineteenth-century United States.[7]

As the industry's structure changed, the price of electricity plummeted—from 4 dollars per kilowatt-hour for U.S. consumers in 1892, to 60 cents in 1930, and just 7 cents in 1970 (in 1993 dollars). Demand soared, doubling every decade in some countries, and the industry adopted a "grow-and-build" strategy in which ever higher levels of demand were counted on to justify scaling up the technology. This, in turn, would lower the price and attract more customers. The advances in power technology during this period stemmed from a surprisingly narrow frontier of advances, however. These incremental improvements included: changes in turbine materials and design, allowing them to operate at higher temperature and pressure; economies of scale in turbines and boilers; and the development of new techniques for cooling the generators.[8]

Following World War II, the United Kingdom still had nearly 600 electricity companies, roughly one-third of them privately owned.

By the sixties, however, this bag of tricks was nearly empty. Average plant efficiencies were leveling off at about 33 percent—meaning that two-thirds of the energy in the fuel was still dissipated as waste heat. Efforts to wring just a bit more out of the technology led to a breakthrough, but not the one expected: For the first time, new power plants were more expensive and less reliable than their predecessors. Yet utility engineers, accustomed to an age of endless progress, resisted the notion that this change was anything but a temporary setback, and continued to argue for ever larger plants.[9]

At about the same time, a sudden plunge into nuclear power plant construction created additional problems for the electric utility industry. Superficially, nuclear plants appeared similar to coal-fired ones; instead of burning fossil fuels to boil water, they split atoms. In the late sixties, utilities ordered scores of nuclear plants, rapidly scaling up reactor technologies originally developed for use in submarines. Soon, the unique hazards of atomic energy were adding dramatically to the complexity and costs of the new plants, raising electricity prices, and generating a public backlash that was heightened by the accidents at Three Mile Island and Chernobyl. By the late eighties, most countries had abandoned their nuclear construction programs and were turning to other power sources.[10]

Matters were complicated by a slowdown in electricity growth in industrial countries, from nearly 8 percent per year during the sixties to an average of 3 percent since the mid-seventies—thanks to higher fuel prices and saturation in the use of some appliances. During this period, the efficiency with which power was used rose, and electricity intensity in industrial countries—measured by the amount of electricity required to produce a dollar of economic activity—leveled off. (See Figure 1.)[11]

Many utility executives saw the trends as a temporary confluence of aberrant economic conditions, perverse government decisions, and an irrational public, and waited expectantly for conditions to return to "normal." When that failed to happen, scores of utilities were stuck with multibillion dollar plants they did not need. Hence, the $30-billion debt of the state-owned national utility in France, in 1994, and the bankruptcies in the eighties of such U.S. utilities as the Washington Public Power Supply System and the Public Service Company of New Hampshire. Electricity forecasting became a guessing game, and utility regulation in the United States was transformed from a dull rubber-stamping process dominated by the industry into a series of confrontations with consumer advocates and government lawyers.[12]

Further complicating the electricity business was the mounting evidence that power plants—particularly the coal-fired behemoths that provide nearly 40 percent of the world's electricity (see Table 1)—cause major environmental problems. Governments passed increasingly strict pollution laws demanding that utilities control everything from coal ash to elusive gases, such as nitrogen oxides. By the early nineties, for example, most new power plants

FIGURE 1

Electricity Intensity of Industrial Economies, 1950-92

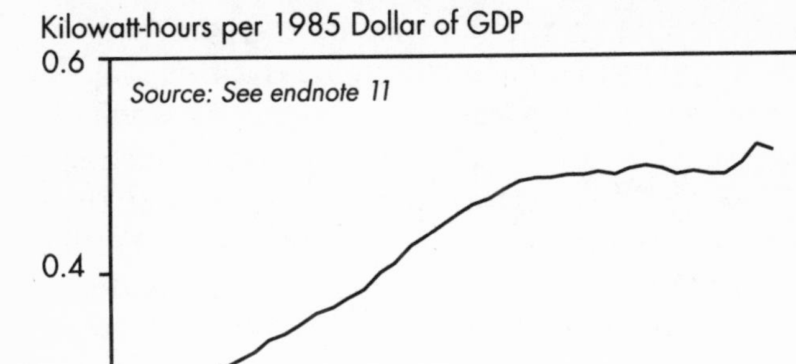

in industrial countries had flue gas desulfurization units to remove the sulfur dioxide that causes acid rain. In some cases, as much as 45 percent of the cost of a new coal-fired power plant stems from environmental compliance. Recently, public concern has also blocked the construction of many plants entirely. These aborted projects include nuclear and coal plants as well as hydroelectric dams opposed by local people concerned about the vast land areas to be inundated.[13]

Electric power systems in developing countries—most of them government-owned—were swept up in the same tidal wave of problems. Pushed by political leaders to cut prices while expanding supply at rates as high as 10 percent annually, managers of many Third World utilities saw both financial stability and service reliability deteriorate. Soaring oil prices and mushrooming debt burdens threw numerous Third World utilities into a period of disarray from which they have yet to recover. On average, rates of

TABLE 1

World Electricity Generation, by Energy Source

Source	1971		1991	
	(terawatt-hours)	(percent of total)	(terawatt-hours)	(percent of total)
Coal	2,142	40.3	4,671	38.8
Renewables[1]	1,241	23.4	2,290	19.0
Nuclear	111	2.1	2,106	17.5
Natural Gas	714	13.5	1,594	13.2
Oil	1,102	20.8	1,376	11.4
Total[2]	5,311	100.0	12,037	100.0

[1]Primarily hydroelectric. [2]Columns may not add to totals due to rounding.

Sources: Organization for Economic Co-operation and Development (OECD), International Energy Agency (IEA), World Energy Outlook (Paris: 1993); OECD, IEA, Energy Statistics and Balances of Non-OECD Countries, 1990-1991 (Paris: 1993).

return on investment fell from 9 percent in the early seventies to 5 percent in the eighties. The faltering performance of some developing countries' utilities has had social and economic repercussions as well. India, for example, experienced "power riots" in the summer of 1993, and government officials say that the country's industrial expansion is being slowed by power shortages.[14]

The World Bank, which plays a lead role in financing electric power projects in developing countries, acknowledged the problems in a 1993 report: "Opaque command and control management of the sector, poorly defined objectives, government interference in daily affairs, and a lack of financial autonomy have affected productive efficiency and institutional performance." In many cases, political manipulation and corruption have made matters even worse. And, given their financial condition, most of these utilities have been unable to add the most basic environmental controls to their power plants.[15]

Still, even with the havoc of recent years, electric utility reform continues to meet heavy resistance in rich and poor nations alike. Utilities provide prosperity for those who control them, and governments often use the power industry to prop up powerful industries. German law, for example, compels the country's utilities to purchase German coal at several times the world price—and at an even greater environmental cost—to protect the jobs of coal miners. This deal is still broadly supported by politicians despite the heavy cost to consumers. In Quebec, the provincial utility subsidizes a costly series of hydroelectric dams; in France, government subsidies in the seventies and eighties created a massive nuclear industry virtually from scratch. Once such industries are in place—and supporting thousands of jobs—pulling the plug on them is no easy political feat.[16]

Despite the powerful interests resisting change, efforts to reform the power industry have gradually gathered momentum since the mid-eighties, driven by an array of economic and environmental forces—and a growing realization that the current system is poorly equipped to deal with today's needs. Scores of utility executives, environmentalists, consumer advocates, and government regulators are calling for new ways of organizing and managing the power industry. Already, some governments have encouraged utilities to contract with independent companies for new power supplies, make sizable investments in improving the energy efficiency of their customers' buildings and equipment, or adopt a new planning strategy, called integrated resource planning, that compares all practical electricity generation and savings options before making investments.

Power industry reformers are also looking to new ways of achieving environmental goals. The old approach—requiring utilities to respond piecemeal to each new pollution law—often leads to a costly series of end-of-the-pipe devices. Today, government agencies are beginning to use the marketplace to encourage the use of cleaner fuels and technologies. Already, 26 U.S. states add indirect environmental costs—the cost of acid rain damage from sulfur and nitrogen oxide emissions, for example—to construction and fuel costs when assessing the competitiveness of proposed new power plants. In some states, such procedures have already tilted the power market away from coal and toward natural gas and renewable resources.[17]

The threat to the earth's climate stability posed by rising concentrations of carbon dioxide further complicates power planning. Since power plants emit 30 percent of the carbon dioxide from fossil fuels, they are central to any effort to reduce the risk of global warming. During the next few years, countries will prepare national climate plans in response to the global treaty signed in Rio de 26 in 1992. Their efforts to improve energy efficiency and deploy technologies to harness renewable energy resources will be vital to achieving climate stability, and are likely to accelerate the process of change in the electric power industry.[18]

Rise of the Independents

One of the most far-reaching developments in recent decades is the emergence of a competitive breed of "independent power producers" (IPPs), who build generating plants and sell electricity to utilities. Until recently, most utility systems were closely guarded monopolies, precluding independent power generation, but today, IPPs are leading builders of new power plants in the United States and several other countries. To the consternation of the defenders of the status quo, this new industry has demonstrated a capacity to construct plants that are smaller, more innovative, and substantially less expensive than their utility brethren. The fresh air of competition has also helped spur a generation of more environmentally sustainable power projects, including many that rely on renewable resources such as wind, solar, and geothermal energy.

Independent power generation has its modern roots in the U.S. Public Utility Regulatory Policies Act (PURPA) of 1978. PURPA allowed independent, unregulated companies to build generating plants that use renewable fuels or "cogenerate" heat for industrial facilities. Intended as a way to boost new technologies, PURPA was implemented unevenly across the United States. The law had its greatest impact in California and Texas, where entrepreneurs began to develop energy projects ranging from large industrial cogeneration facilities to wind farms and wood-fired power plants.[19]

As of 1992, the United States had 55,000 megawatts of independent generators in operation—7 percent of the nation's total generating capacity. The bulk of these generators—totalling 39,000 megawatts—had been added since 1980. In 1992, independents, for the first time, added more generating capacity in the U.S. than did utilities. (See Figure 2.) Many experts expect this trend to accelerate, in part because the Energy Policy Act of 1992 further liberalizes the power industry—requiring readier access to utility lines, and encouraging more utilities to form IPP subsidiaries. Already, a growing number of states are holding regular power auctions in which competing independent suppliers bid on contracts to provide electricity to utilities. In a further step, electricity contracts are expected to be traded on U.S. commodities markets in much the way oil, soybeans, and many other commodities already are, beginning probably in 1995.[20]

The fresh air of competition has helped spur a generation of more environmentally sustainable power plants.

The independents are also spearheading a shift to smaller generators in many countries. Power plants owned by independent producers in the United States averaged just 25 megawatts in 1992, one-fourth the size of the average utility plant. Spurred in part by growing competition, utilities, too, are starting to build smaller facilities. The average size of U.S. utility-built plants declined from more than 600 megawatts in the mid-eighties, to an average of about 100 megawatts in 1992. The new, smaller generators range from gas turbines to wind turbines, and from geothermal power plants to solar cells. Only nuclear power, with its roots in centralized government and utility control, has been bypassed by this renaissance of entrepreneurialism in the power industry.[21]

Among the countries fostering an extensive IPP industry are India, the Netherlands, and the United Kingdom. The United

FIGURE 2

Additions to U.S. Generating Capacity by Electric Utilities and Independent Power Producers, 1950-92

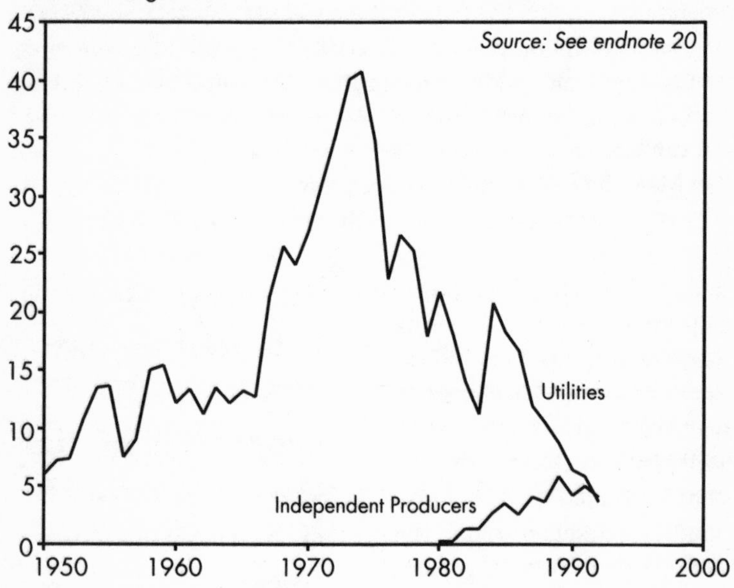

Kingdom currently has one of the world's most wide-open power systems, which, in the space of just a few years, has largely destroyed the country's coal industry and created a boom in gas-fired power plants. Several other nations—including Denmark, Germany, Japan, and Switzerland—allow some independent power generators, mainly as a way to encourage small wind and solar projects. Many developing countries burdened by severe financial constraints and power shortages—including China, Indonesia, Mexico, Pakistan, and Syria—have also turned to private companies to build and operate new power plants.[22]

Although IPPs are proliferating, they remain the exception rather than the rule. In most nations, in most circumstances, it is still illegal for any company that is not a regulated utility to feed electricity into the power grid. In Europe, however, the European Union has begun to challenge these monopolies as a violation of the free trade laws that now govern commerce in most of western Europe. Such reforms are fiercely resisted in several countries, particularly in France, where the national power monopoly is the lifeline to the country's large nuclear industry, and in Germany, where coal from the Ruhr Valley is protected via a $4-billion a year subsidy paid by all German electricity consumers. These monopolies will not die easily or quickly, but, in the long-run, the European power industry appears headed toward a far more open and competitive market.[23]

Major industrial users of energy were among the first to take advantage of the opportunity to sell electricity on the grid. Most of these companies require large amounts of heat—which they produce on site—as well as electricity in their production processes. One way to reduce a factory's energy bills is to generate electricity within the factory itself, and use the waste heat from that process for other applications within the plant. This process (known as combined-heat-and-power or cogeneration) allows up to 90 percent of the energy in a fuel to be used for productive purposes—far above the 33-percent average for the central power plants owned by U.S. utilities. As a rule of thumb, the value of the hot steam produced by a cogenerator knocks the equivalent of about 1-2 cents per kilowatt-hour (15-30 percent) off the cost of power generation.[24]

Although cogeneration has been common in some industries for decades, it was not until the reform of utility laws in the late seventies and early eighties that cogeneration gained a second wind. In fact, U.S. industry increased its cogeneration capacity from 10,500 megawatts in 1979 to 40,700 megawatts in 1992. During that period, the development of smaller generators has allowed even modestly sized commercial establishments—such as hotels, hospitals, and fast food restaurants—to cogenerate power at a lower price than that offered by utilities. Some companies offer factory-built "packaged cogeneration" systems, based on adapted diesel engines that provide as little as 5 kilowatts of electricity. This approach has been especially popular in Denmark, where some 900 megawatts-worth of small, gas-fueled cogenerating plants are operating or on order at local district heating plants.[25]

The past decade has also witnessed the rise of a new breed of independent power company that is ready to build nearly any kind of generating plant anywhere in the world. These companies range from small firms such as AES or Kenetech to multinational corporations such as ABB and Texaco. In addition, by the mid-nineties, most major U.S. utilities had IPP subsidiaries, usually with catchy regional names such as Mission Energy in southern California, and Dominion Energy in Virginia. Unlike traditional utilities, these new producers operate as unregulated companies: They sign multi-year contracts with utilities, locking in the price they will be paid for their power, then raise their own capital—at a price reflecting the perceived risk of the project—and assume the remaining risk of cost overruns or delays. Although many utility experts had predicted such a system would never work, the success of independent power probably would not surprise Thomas Edison—or Adam Smith, for that matter.[26]

A host of innovative technology has accompanied the rise of the independent power industry. One of the first barriers to fall was the one that had confounded utility engineers since the sixties: the inability to build reliable power plants with efficiencies much higher than 40 percent. That limit has been easily surpassed by using a surprisingly familiar device—the jet engine. A jet engine burns a pressurized mixture of fuel and air to spin a turbine that generates thrust and propels the plane forward. The same basic device can be readily converted to stationary use as a mechanical power generator—usually fueled by natural gas or diesel oil—a configuration known as a gas turbine. Beginning in the sixties, electric utilities employed gas turbines as a way to meet relatively brief "peaks" in power demand. Early gas turbines were less than 30 percent efficient, however, and rising fuel prices in the seventies led to a hiatus in their use.[27]

It was not until the late eighties that power companies—mainly independents—again turned their attention to generating plants based on gas turbines. This renaissance focused on natural gas-fueled "combined-cycle plants," in which excess heat from a gas turbine powers a second steam turbine (similar to those used in conventional power plants). This configuration achieved an efficiency of 40 percent by the late eighties, rising to 50 percent for a General Electric plant opened in South Korea in 1993, and 53 percent for a new design announced by ABB that same year. Already, these combined cycle systems have far surpassed the maximum efficiency of simple steam-cycle power plants burning fossil fuels. (See Figure 3.) Still, engineers expect continuing improvements during the next decade.[28]

Such plants have other advantages. At roughly $700 per kilowatt—just above half the cost of a conventional coal plant—they are inexpensive to build and can be brought on line quickly: two and a half years for the huge, 1,875-megawatt Teeside cogeneration plant in the United Kingdom. This is roughly half as long as a similar-sized coal plant would take to complete. The difference is largely because combined-cycle plants are built from modular, factory-built units; for example, Teeside is comprised of eight gas turbines and two steam turbines.[29]

Smaller, more versatile turbines are also on the way. Borrowing from the advanced metals, new blade designs, and high compression ratios of today's jet engines (which have benefitted from billions of dollars worth of government-funded research and development), several companies are now producing smaller "aeroderivative" turbines. By applying a host of innovations, such as steam injection, the efficiency of these devices has reached 39 percent and is expected one day to approach 60 percent. Since aero turbines are factory-built, they cost as little as $350 per kilowatt, and engineers can install them in just a few months. Moreover, because of their size—48 megawatts from the largest commercial aircraft engine, and 1 megawatt or less from smaller models—aircraft turbines suit a wide range of applications. A large apartment building could use one as a cogenerator, for instance.[30]

Combined-cycle turbines and aeroderivative turbines also have environmental advantages over conventional oil- or coal-fired generators. When burning natural gas, they emit virtually no sulfur and only negligible particulates. They also cut emissions of nitrogen oxides (NO_x) by up to 90 percent and carbon dioxide emissions by up to 60 percent compared to conventional plants. (See Table 2.) Indeed, the combination of low cost

FIGURE 3

Electrical Efficiency of Fossil Fuel Power Plants, 1882-1993

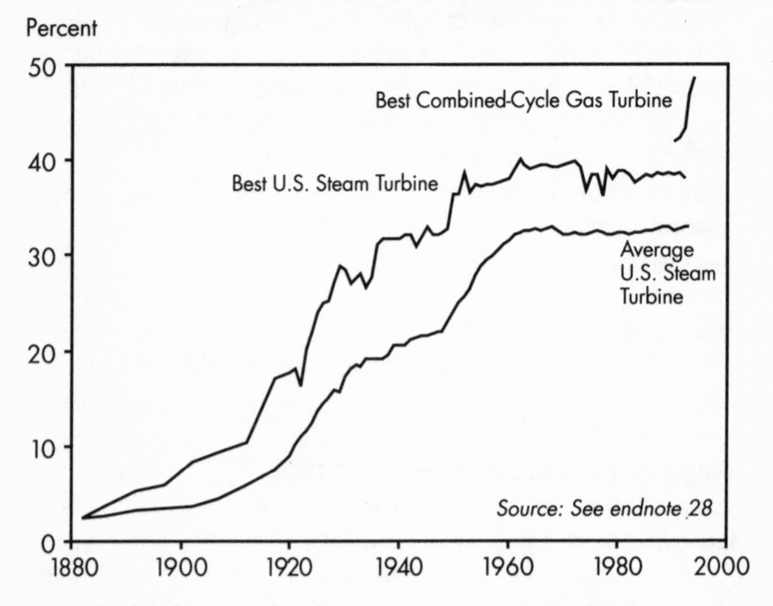

and low emissions could spur utilities to convert hundreds of aging coal plants into gas-burning, combined-cycle plants—for as little as $350 per kilowatt. And from the Tennessee Valley in North America to Ukraine in eastern Europe, studies recommend converting partially completed nuclear plants into combined-cycle models at less than the cost of finishing them as nuclear plants.[31]

Worldwide, some 400,000 megawatts of gas turbine plants could be built by 2005, according to a General Electric forecast: Units are already up and running in countries as diverse as Austria, Egypt, Japan, and Nigeria. A secondary result of this boom is the recent emergence of natural gas as the dominant fuel for new power plants in many countries.[32]

Although cleanest when fueled by natural gas, the new turbines can burn a variety of fuels, including coal, if it is first gasified by rapid heating in an oxygen-depleted atmosphere. A few

TABLE 2

Conversion Efficiencies and Air Pollutants, Various Electricity-Generating Technologies[1]

Technology	Conversion Efficiency	Emissions NO_x	SO_2	CO_2
	(percent)	(grams per kilowatt-hour)		
Coal[2]				
Conventional Steam Cycle (without scrubbers)	36	1.29	17.2	884
Conventional Steam Cycle (with scrubbers)	36	1.29	0.86	884
Fluidized Bed	37	0.42	0.84	861
Integrated Gasification Combined Cycle	42	0.11	0.30	758
Natural Gas				
Fuel Cell	36	0.04	0.00	509
Aeroderivative Gas Turbine	39	0.23	0.00	470
Combined-Cycle Gas Turbine	53	0.10	0.00	345

[1]Data are for particular plants representative of those in operation or under development. [2]Coal plants are burning coal with 2.2 percent sulfur content.

Source: See endnote 26

small coal gasification plants have been tested successfully, and larger units are under construction in the Netherlands, Spain, and the United States. Still, the market prospects for this technology are clouded. While coal gasification plants achieve 85 percent lower sulfur emissions compared to conventional units, carbon emissions drop only 15 percent. Moreover, coal gasification is expensive compared to a simple gas turbine plant.[33]

The same technology can also be adapted to run on biomass fuels such as wood or agricultural wastes, or even specially grown energy crops. The first prototype—a six megawatt wood-fired, combined-cycle turbine—was commissioned at a district heating plant in Värnamo, Sweden, in 1994. Bahia, Brazil, is planning a 25-megawatt, wood-fired plant with the support of the Global

Environment Facility. A scaled-up version of the Värnamo combined-cycle gas turbine is expected to convert more than 45 percent of the wood's energy to usable electricity.[34]

Biomass-fueled power plants' environmental advantages include the lack of sulfur emissions, and, so long as an equivalent amount of biomass is regrown, no net carbon emissions, either. Although biomass provides only about one percent of the world's electricity today, it already plays a much larger role in several countries, including Finland, Mauritius, and Denmark. According to one estimate, today's wood and agricultural residues could provide as much as 30 percent of the world's current electricity supply—without damaging forests or agriculture.[35]

Earth, Wind, and Fire

Three other virtually inexhaustible resources have recently emerged as economically competitive means of generating electricity. Geothermal, wind, and solar energy produce limited if any air pollution, and could one day provide most if not all of the world's electricity. Since the early eighties, some 12,000 megawatts of power plants fueled by these three sources as well as biomass have been installed in the United States, more than 70 percent of which is owned by independents. In California, where half the development has occurred, these four "new renewables" provided 11 percent of the state's electricity in 1993.[36]

Geothermal energy comes from the earth's heated core—the power behind earthquakes and volcanoes. First used to generate electricity in Italy in 1904, geothermal power generation is now common in some regions. In areas such as The Geysers in California, a well-placed borehole can bring large amounts of hot steam to the surface, which can then power a turbine and generator, just as in a traditional thermal power plant. Worldwide geothermal-generating capacity was estimated at 7,000 megawatts in 1993. It provides 28 percent of the power in Nicaragua, 26 percent in the Philippines, and 9 percent in Kenya.[37]

By the turn of the century, world geothermal generation is expected to grow to 15,000 megawatts, installed in some 40 countries. In the United States, for example, the Department of Energy (DOE) estimates that hydrothermal reservoirs—hot water or steam trapped in rock fractures much the way oil and gas are—could in theory provide 30 times as much energy as the country currently uses. Japan, which currently has 270 megawatts of geothermal capacity installed, has an estimated potential of more than 69,000 megawatts, nearly double the country's current nuclear electric capacity. Other countries, including Djibouti and Saint Lucia, could reasonably produce most of their electricity from geothermal resources that have already been identified. Although geothermal reserves can be depleted if managed incorrectly (and some have been), worldwide resources are sufficiently large for this energy resource to be treated as renewable.[38]

Geothermal, wind and solar energy could one day provide most of the world's electricity.

Economics remains the major constraint to greater reliance on geothermal energy. The cost of tapping geothermal heat is currently too high for all but the most concentrated resources, and it is unlikely to fall as dramatically as the cost of manufactured technologies such as wind and solar generators. Nonetheless, geothermal energy has the advantage of providing steady, "baseload" power, much the way that conventional coal plants do. For a growing number of countries, geothermal energy will likely be an important power source in the decades ahead.

Wind-generated electricity is also joining the ranks of commercial power sources. By the mid-nineties, following a decade and a half of steady technical improvements, nearly 20,000 turbines had been installed worldwide, producing 3,000 megawatts of electricity. In Denmark, the wind provided 3 percent of the country's power in 1993, and in California, 1.2 percent. Some 5,000-10,000 megawatts of wind power are projected to be added by the year 2000—most of it in northern Europe and the North American Great Plains and Midwest. Sizable wind projects are also on the drawing boards in Argentina, China, India, Mexico, New Zealand, Spain, and other countries.[39]

By employing larger turbines, advanced blades, electronic controls, and improved transmissions and generators, wind manufacturers have already lowered costs to 5-7 cents per kilowatt-hour, which is competitive with fossil fuel-based power in many regions. Yet this technology is still in its infancy. As advances proceed and mass production begins, further cost declines will make wind power one of the world's least-expensive sources of electricity.[40]

Relying heavily on wind power will inevitably raise land-use issues. While wind farms would "occupy" large areas, most would be land with little vegetation and even fewer people or animals. Moreover, wind machines primarily occupy land in a visual sense. The area surrounding the turbines can be used as before—usually for grazing animals or raising crops—while providing farmers with supplementary income. In Wyoming, for instance, a hectare of rangeland that sells for $100 could yield more than $25,000 worth of electricity annually. In many windy regions, harnessing the wind might enhance land values by interrupting wind flow thereby reducing erosion. However, some wind farms have killed birds, a problem that needs to be solved if wind energy is to realize its potential.[41]

Even excluding environmentally sensitive areas, the global wind energy potential is roughly five times current world electricity use. Since the power available from wind rises with the cube of the wind speed, most of the development will occur in the very windiest areas. In the United States, where detailed surveys have been conducted, it appears that wind turbines installed on 0.6 percent of the land area of the lower 48 states—mainly in the Great Plains—could meet 20 percent of current U.S. power needs. Indeed, conservative resource estimates show that even if large environmentally sensitive areas were excluded, three U.S. states—North and South Dakota and Texas—could in principle supply all the country's electricity. Although no one expects such a scheme to be implemented, it is clear that wind power could become a major component of North America's electric grid.[42]

Many other nations are rich enough in wind resources that they could in theory derive all their electricity from the wind. They include Argentina, Canada, Chile, China, and Russia. Others—including Egypt, India, Mexico, Tunisia, and South Africa—could easily push reliance on wind power to 20 percent or more. Europe as a whole could obtain between 7 and 26 percent of its power from the wind, depending on how much land is excluded for environmental reasons. At least 20 small, subtropical island countries have nearly constant trade winds that could meet a large share of their electricity needs. In addition, many regions with limited winds may find that the limits end at the water's edge: New England, the United Kingdom, and Poland are among those that could obtain huge amounts of power from wind farms located on offshore platforms in shallow seas.[43]

The global wind energy potential is roughly five times current world electricity uses.

Solar energy, the most abundant energy source of all, is about a decade behind wind power in terms of commercial development, but has already shown strong promise as a source of utility power. The most cost-effective, large-scale solar technology is the solar thermal power plant. In a solar thermal plant, mirrors concentrate the sun's rays to heat a liquid, producing steam for an electricity-generating turbine. Solar thermal electric systems generally require direct sunlight, limiting them to arid and semi-arid regions.

Between 1984 and 1990, Luz International installed more than 350 megawatts of solar thermal generating capacity in California's Mojave Desert. Spread over 750 hectares, the parabolic trough collectors produce enough power for about 170,000 homes. And they do so for as little as 9 cents per kilowatt-hour, far below the 26-cent cost at Luz's first plant in 1984. By using natural gas to keep the turbines running when the sun is not shining, the company provides Southern California Edison with power whenever it is needed. Although Luz was forced into bankruptcy in 1991, its 9 plants had been transferred to separate owners and continue to churn out power. Meanwhile, investors purchased rights to the technology from Luz's creditors, naming the reorganized, Israel-based company, Solel. As of early 1994, Solel was working to raise the efficiency and improve the storage of its trough systems, and negotiating to build one or more 200-megawatt solar power plants—possibly in Brazil, India, Israel, or the United States. The Global Environment Facility, the funding mechanism for the Framework Convention on Climate Change, has shown interest in supporting this technology in developing countries.[44]

Unlike parabolic troughs, which track the sun along one axis, parabolic dish collectors follow the sun along two axes: They focus sunlight onto a single point where heat can be either converted directly to electricity or transferred by pipe to a central turbine. The dishes are generally more thermally efficient than troughs. Both types of equipment are built in moderately sized, standardized units, allowing generating capacity to be added incrementally as needed. The U.S. Department of Energy expects parabolic dishes to produce power for 6 cents per kilowatt-hour early in the next century, and some experts believe that such systems will outperform troughs economically. An initial 2-megawatt plant was under construction in 1994 in Australia's remote Northern Territory. The project is funded by a consortium of electric utilities under pressure to reduce their reliance on coal.[45]

Sunlight can generate electricity by other means as well. Some utility engineers favor the "power tower": a tall structure bearing a 10-200 megawatt solar receiver; mirrors surrounding the tower focus sunlight on it. The technology's chief advantage is a heat storage system using molten salt that allows power generation after the sun has set. But towers would likely be up to 200 meters tall, a centralized configuration that may raise environmental objections. Another technology, the solar photovoltaic cell, is the most versatile of solar generators, and as discussed later, may find its main role within buildings.[46]

With solar power, the adequacy of the resource is hardly an issue. The amount of solar energy reaching the earth's surface each year totals more than 10,000 times the electricity used by all human beings each year. To provide as much electricity as the world used each year would require installing solar thermal power plants on a land area equivalent to that of the small nation of Panama—or less than one-quarter of the U.S. state of Arizona. At the national level, all U.S. power needs could be met with solar plants spread over 20,000 square kilometers, roughly one-tenth the area occupied by domestic U.S. military facilities.[47]

The largest challenge of intermittently operating solar power plants or wind turbines is integrating them into the grid. In the past, utility engineers have argued that fluctuating energy sources would create havoc in their systems, and require costly investment in backup generators or storage devices. But experience to date with wind generators in California and elsewhere suggests otherwise. Wind power has been easily integrated into the existing mix of generators—reaching as high as 20 percent of the total in some regions—and has actually increased the reliability of some systems.

With the help of new electronic controls, renewables can improve reliability and cost-effectiveness of many utility systems.

Still, utility engineers must deal with the fact that intermittent renewable power sources do not fit within the traditional hierarchy of generators. Most of today's power systems use a combination of "baseload" (usually coal or nuclear) plants that operate most of the time, "intermediate" plants that turn off at night, and "peaking" units that operate only when demand is highest (usually gas turbines or hydro plants). Solar and wind generators do not fit neatly into any of these roles, but the challenge is not qualitatively different from one that utilities mastered long ago: meeting the rapidly fluctuating demands of customers. The experience garnered by California's Pacific Gas and Electric (PG&E) and Southern California Edison indicates that a diverse array of intermittent power sources can meet one third of a utility's load at no additional cost, and up to one-half at an additional cost of only 10 percent.[48]

The ease of integrating renewable power sources depends in part on how well their availability matches patterns of consumer demand. Experience shows, for example, that while in some regions peak winds coincide nicely with peak power demand, in others they do not. The winds of the northern Great Plains coincide with utilities' winter peak loads, and, in northern California's Altamont Pass, there is a good but not perfect match with peak summer demand. In the summer, peak demand in California comes between 2:00 p.m. and 8:00 p.m., while the highest winds occur between 5:00 p.m. and midnight. However, if coupled with solar generators that peak earlier in the day, wind and solar power together would fit the power use patterns in northern California almost perfectly.[49]

If intermittent power sources are to supply 20 percent or more of a region's electricity, adjustments may be needed. In regions with extensive hydropower, such as the northwestern United States, little if any additional backup is required since hydro already provides a reserve supply. In addition, the new generation of gas turbines described earlier are sufficiently inexpensive that they make economic sense even if operated partly on standby, raising the possibility that in the future, independent power producers might build combination gas-turbine and wind-turbine power plants. Together they would provide high reliability and low installation and operating costs.[50]

As reliance on renewable energy sources grows, additional electricity storage may be needed in some regions. The most commercially ready alternative is "pumped hydro storage," in which excess electricity produced during periods of low demand is used to pump water up to a reservoir, from which it is released to generate power at times when demand is high. Another alternative that is economically feasible in many areas is compressed air storage in which underground rock fractures are filled with high pressure air that can be used to help spin a gas turbine during times of peak demand. A third storage alternative is unique to solar thermal projects. Heat produced during the day can be used to heat rocks, water, or another medium, which can then be used to keep the system's turbines running long after the sun has set. Studies show that heat storage can extend the operating period of such a plant by several hours at only minor additional cost.[51]

Although intermittent renewables will continue to present challenges for utility engineers, most are amenable to relatively simple solutions. In the end, with the help of new electronic controls, renewables are likely to improve both the reliability and cost-effectiveness of many utility systems. As David Mills, a solar energy expert at Australia's University of Sydney, puts it: "Despite the tenacious myth that baseload is beautiful, the ideal power plant would produce an output...as flexible as possible...accounting for diurnal variations in demand as well as changes in the availability of other generating plants on the utility grid." This level of performance is beyond that of traditional utility generators, but is well within reach of the solar power plants being designed by Mills and other scientists.[52]

The combination of a growing independent power industry and the emergence of new generating options suggests that, beginning almost immediately, electricity supplies will be less expensive *and* less environmentally damaging than in the past. (See Table 3.) Moreover, the small scale and diverse nature of these power sources will reduce the risk of cost overruns.[53]

This shift was highlighted by a power auction held in California in 1993. Independent producers offered several thousand megawatts of bids, generally at prices well below those of proposed utility projects, and also below the prices of the last generation of independent plants built in the late eighties. In addition, this auction, which included a separate set-aside for renewable projects, suggested that wind and geothermal energy can now provide power at an average cost of just 5 cents per kilowatt-hour, compared to an average of about 4 cents per kilowatt-hour for gas-fired plants. If this momentum continues, the way may be paved for a far richer array of power options in the near future.[54]

TABLE 3

Cost of Electric Power Generation in the United States, 1985, 1994, and 2000

Technology	1985	1994	2000
		(1993 cents per kilowatt hour)	
Natural Gas	10-13	4-5	3-4
Coal	8-10	5-6	4-5
Wind	10-13	5-7	4-5
Solar Thermal[1]	13-26	8-10	5-6
Nuclear	10-21	10-21	*[2]

[1]With natural gas as backup fuel [2]No plant ordered since 1978; all orders since 1973 subsequently cancelled.

Source: See endnote 46

The Demand-Side Revolution

In the mid-seventies, a maverick group of energy analysts reached a startling conclusion: Both electric utilities and their customers might be better off if they invested in ways to reduce power use. Studies showed that improving the efficiency with which electricity is used often costs less than building and operating new plants. When lawyers for the Environmental Defense Fund and other groups brought this argument to public hearings in the United States, utilities fiercely resisted, accustomed as they were to promoting power use, often by charging less to those who use more. One campaign in the fifties featured Ronald Reagan touting all-electric "Gold Medallion" homes.[55]

The environmentalists lost some early battles, but regulatory commissions eventually ordered utilities to adopt efficiency programs. Early efforts were modest: eliminating discounts for big power users, adding "bill-stuffers" that urged a lowering of thermostats, and offering free energy audits to customers. But the efforts gradually expanded to include cash rebates for efficient appliances or compact fluorescent lamps, low-interest loans for home weatherization or industrial retrofits, and even rebates for the purchase of solar water heaters. The new approach became known as demand-side management (DSM) and more colloquially as "negawatts." In the United States, DSM is now a major part of the business for many utilities—absorbing more than $2 billion annually in utility investments in recent years.[56]

At the heart of the DSM revolution is a challenge to the utility axiom that the more electricity consumed, the better. According to the new thinking, electric power is an expensive and highly directed form of energy, and using it to provide low-grade heat is inherently wasteful. Physicist Amory Lovins, for example, compares heating water with electricity to "cutting butter with a chain saw." He and other analysts point out that electricity now provides just 17 percent of the end-use energy of industrial countries, a figure that need not rise much further—despite increasing use of electronic equipment—if power is used more efficiently.[57]

Although the notion that it costs less to save electricity than to supply it is straightforward, the idea that utilities themselves should discourage the use of their product is counterintuitive. But then, utilities do not operate by standard rules of the market. They are chartered and regulated by governments, face minimal financial risks, and have access to capital at low interest rates—allowing them to invest in power plants with a payback period of 20 years. In the past, those advantages have served as a subsidy for generating electricity, while capital- and information-short consumers—with a two-to-three year payback horizon—shoulder the full responsibility for investments in efficiency. The result is predictable: Customers end up paying for unnecessary power plants that send current to outdated, inefficient appliances.

The first attempts to remedy this imbalance in the eighties penalized utilities that failed to invest in conservation. This "big stick" approach worked in several states, but as the eighties drew to a close, investments in efficiency fell as executives saw that they were losing money on DSM. The reason: U.S. utilities have traditionally been regulated so that profits are linked to electricity sales. Anything that lowers power use therefore reduces the return to investors, even if the efficiency investment is added to the capital base on which regulators allow utilities to earn profits. Such a system places the economic interests of electricity consumers at odds with the interests of the shareholders to whom utility executives are responsible.[58]

The solution to this problem was pioneered by enterprising utility executives, regulators, and environmental advocates, who realized, in the words of New England Electric System President John Rowe, that "the rat must smell the cheese." In 1989, the National Association of Regulatory Utility Commissioners proposed that regulators offer "cheese" by compensating utilities in a variety of ways for profits that would be lost from reduced sales. In some cases, companies are now allowed to earn higher returns for efficiency investments than for building power plants. In others, utilities gain a share of the money saved. Such approaches are spreading rapidly in the United States, and are being experimented with in the United Kingdom.[59]

Beyond adjusting incentives, 30 states have ordered utilities to adopt a new approach to charting their future resource mix called integrated resource planning (IRP). Using IRP, planners assess the benefits, risks, and costs—sometimes including environmental costs—of all practical generation and savings options. The winning option may be determined by seeking competitive bids from energy-saving companies as well as power suppliers. Sometimes even an efficiency program that raises power prices may be advantageous, so long as it cuts average consumption and thereby lowers customers' monthly bills. IRP also prompts utilities to look at how they can improve the efficiency of transmission and distribution of energy. Newly developed transformers, for example, use 70-90 percent less electricity than conventional ones, a savings magnified by the fact that virtually all electric power passes through at least two transformers between the generating plant and the consumer. For many utilities, IRP has spurred a reassessment of the very nature of their business: They start to see themselves less as commodity suppliers and more as full-service energy companies—much as Edison envisioned 11 decades ago.[60]

In response to such policies, U.S. utility spending on efficiency rocketed from less than $900 million in 1989 to an estimated $2.8 billion in 1993. (See Figure 4.) Most of this investment is concentrated in California, the Northwest, and the Northeast, where state regulators (and the boards of publicly owned systems) have been most supportive. California's Public Utilities Commission estimates that utility efficiency investments made between 1990 and 1993 will yield some $1.9-billion worth of net benefits over the next few decades. A 1993 survey by the Electric Power Research Institute, a utility-funded organization in Palo Alto, California, of most U.S. DSM programs found that they saved electricity at an average cost of just 2.1 cents per kilowatt-hour. Other studies, however, indicate that poorly managed DSM can cost several times this much. Careful design and implementation of DSM is needed, including rigorous evaluations and the flexibility to redirect programs operating below par.[61]

A 1993 survey found that utility DSM programs save electricity at an average cost of 2.1 cents per kilowatt-hour.

Utilities in Canada and western Europe are also adopting demand-side management. Two of Canada's provincial utilities, Ontario Hydro and BC Hydro, have large programs, while a third, Hydro-Québec, launched a sizable DSM program in 1994. More than 50 European utilities started lighting efficiency programs between 1987 and 1992; lighting DSM programs now exist in 11 nations. Exceptionally strong efforts are found in municipally owned power companies in Denmark, Sweden, and the Netherlands, with Dutch utilities spending around $30 million each year on lighting alone. Data from 40 of these efforts indicate that, even with administrative expenses included, improving the efficiency of lighting costs on average just 2 cents per kilowatt-hour, far less than producing power.[62]

FIGURE 4

U.S. Utility Spending on Demand-Side Management, 1989-93

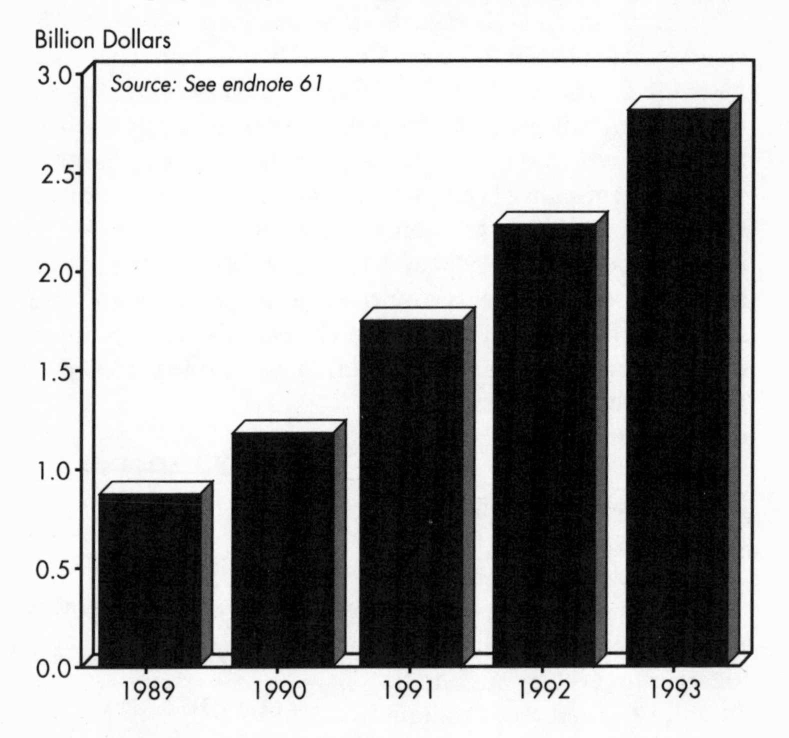

Though their programs continue to grow, European utilities still lag about a decade behind their U.S. counterparts in their cultural acceptance of DSM, according to Evan Mills, a scientist at Lawrence Berkeley Laboratory in California. One explanation is that European power companies typically run DSM on the sidelines rather than as an integral part of their businesses. In the Netherlands, for example, lighting efficiency programs are funded by a tax on electricity, rather than by the utilities directly. In fact, most European programs are still born of simple government mandates, and are not set up to be profitable. The giant German utility, RWE, for example, agreed to spend 100 million deutsche marks ($60 million) on DSM between 1993 and 1995 as part of a bargain with the government that will allow the company to open a low-grade coal mine to fuel its power plants.[63]

DSM has begun to catch on in other parts of the world as well. In Thailand, where double-digit power growth is nearly bursting the seams of the energy infrastructure, the national utility has launched a five-year, $190-million program that includes purchasing efficient lights, appliances, and motors. Brazil, China, and Mexico are among other nations developing DSM programs. The Lawrence Berkeley Laboratory estimates that efficiency could cut the growth of power use in developing countries by 25 percent over the next three decades, freeing up billions of dollars for other investments. Brazil alone could reduce the projected growth in electricity use in 2010 by 42 percent, estimates Howard Geller of the American Council for an Energy-Efficient Economy. And the transitional economies of Eastern Europe and the former Soviet Union could probably meet all growth in their electricity needs through 2010 with today's generating capacity just by improving their dismally low efficiency levels.[64]

By the early nineties, utility efficiency programs had proven their ability to cost-effectively limit power use. The Netherlands expects to cut power use by 2.5 percent by the end of the decade via lighting programs alone, and U.S. utilities anticipate that electricity sales in 2000 will be 4 percent lower than previously projected, and in 2010, 6 percent lower. The most aggressive U.S. utilities are planning to reduce sales by more than 6 percent by 2000, with DSM offsetting half their projected growth in demand. And the largest privately owned U.S. power company, Pacific Gas and Electric, aims to meet three-quarters of its projected rise in demand for energy services via efficiency. As a result of DSM and other efficiency programs, California held per capita power use in 1993 to the 1979 level, while it rose 22 percent in the rest of the United States. (See Figure 5.) If all utilities avoided just half their projected growth through efficiency, U.S. electricity use in 2010 would be 19 percent lower, according to Eric Hirst of Oak Ridge National Laboratory in Tennessee.[65]

Demand-side management programs also deliver other benefits. Since DSM has created a multibillion dollar market in more efficient technologies, it has encouraged manufacturers to invest in improved light bulbs and super-insulating windows.

FIGURE 5

Per Capita Electricity Use in California and Rest of United States, 1960-93

Thousand Kilowatt-hours
Per Person

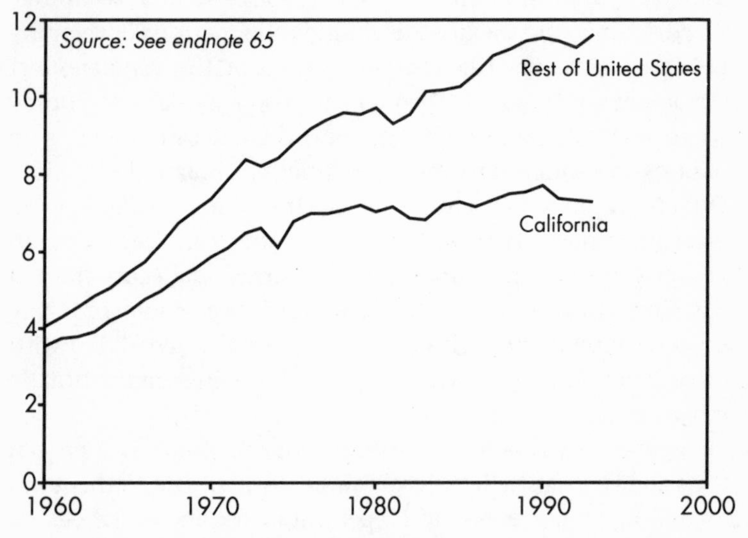

Some utilities have banded together to create incentives for manufacturers to produce less electricity-hungry products. In one such effort, 24 U.S. utilities established a $30-million prize for an efficient refrigerator competition. The winning entry—manufactured by Whirlpool—slashes electricity use by 30 percent compared with commercial models of an equivalent size. Utility DSM programs have also led to the development of a whole new industry of energy service companies. These firms play the same role that independent power producers do in the generation business, packaging investments in a range of energy-saving measures in factories, businesses, and homes, selling "saved energy" to utilities.[66]

DSM programs are off to a strong start in North America, but they have a long way to go to realize their potential. The greatest challenge may be persuading governments to require utilities to adopt full IRP that considers environmental costs as well as direct investments, and provides strong financial incentives to invest in efficiency. For developing countries, a top priority is to change the lending priorities of the World Bank and regional development banks, traditionally hostile to efficiency investments. Throughout the eighties, their lending for new power plants outstripped that for end-use efficiency by about 100 to 1. And although the World Bank approved a policy intended to correct this imbalance in 1992, only 2 of the 46 electricity loans being considered in the first half of 1993 were consistent with the new approach. The Asian Development Bank decided in 1993 to require environmental cost assessments of new power projects; it remains to be seen how effectively the new policy is implemented.[67]

Growing investment in demand-side management raises broader questions about electricity's role in the economy. First, how much more efficiently might electricity be used in existing applications? Second, will the resulting reductions in power use be offset by major new uses of electricity? Since the mid-seventies, power use in most countries has grown two to three times as rapidly as energy use as a whole, and some utility industry analysts expect this trend to continue indefinitely. However, given the relatively high cost of electricity compared to other forms of energy—particularly natural gas—and the fact that electricity is beginning to saturate its major markets in industrial countries, growth in electricity use seems likely to level off. In fact, recent data in the United States suggest that it is already beginning to do so. Although new technologies such as computers and electric-arc steel furnaces will continue to develop, these could easily be offset by more efficient lighting and appliances, and reduced use of electricity in wasteful applications such as home heating.[68]

There is, however, one new use of electricity that may have a larger effect on utilities. Companies around the world are at work on electric cars, trucks, and buses, spurred in part by strict new clean air standards in California and other regions. Whether the electric car will one day make the internal combustion engine obsolete is unclear, and will hinge in part on the development of an inexpensive, long-range battery. But even though they would not add greatly to projected power demand—half the

U.S. auto fleet could be converted to electrics, while increasing the country's power supply by only 10 percent—electric cars could affect the power industry in other ways. Drivers would likely recharge such cars at home, and utilities might help install the recharging devices, probably in conjunction with load-management controls that ensure that the recharging occurs during off-peak periods. California utilities have asked regulators to allow them to use their power revenues to subsidize installation of recharging equipment and even customers' purchase of electric cars. In short, the definition of a "full-service utility" seems destined to continue expanding for some time.[69]

Decentralizing the Power System

The traditional electric utility model—and the structure of the industry—is based on the assumption that large, central stations are the most economical way to provide power to customers. Yet the advent of new, smaller gas turbines and renewable power generators have shaken this conventional wisdom. Soon the traditional model may be obliterated entirely by fuel cells mounted in basements, and rooftop solar systems that allow residential customers to generate their own power and sell excess supplies to other users through the grid.

The move to a power system that relies on a broad mix of large and small generating plants could simultaneously improve efficiency and lower the environmental burden of today's electric power systems. It would also reduce the need to build and upgrade transmission and distribution lines. A typical company with 50 power plants connected to its system today could see the figure reach 5,000 or even 50,000 by 2010. The change would be similar to that of a corporation that used 3 mainframe computers in 1980, and 30,000 personal computers in 1994. In both cases, this shift requires major changes in the way the system is operated.

One of the most revolutionary technologies is the fuel cell, which produces electricity without using the mechanical generators that produce most power today. Invented in the early nineteenth century, the fuel cell is an electrochemical device, consisting of an electrolyte and two electrodes. It operates much the way a chemical battery does, creating an electric charge by capturing the electrons temporarily released when hydrogen and oxygen ions are combined to make water. Unlike a battery, however, fuel cells require constant fueling—usually hydrogen or methane—and last far longer than most batteries do. The technology came into its own during the sixties: Space scientists looking for a small, self-contained power system rediscovered the fuel cell and invested billions of dollars developing it for use in spacecraft. Today, fuel cells generate most of the electricity for the U.S. space shuttle when it is in orbit.[70]

The fuel cell has three major advantages over conventional power generators: It is relatively efficient, converting 40 to 70 percent of the energy potential of fuels into electricity. Since it lacks a conventional generator, it produces minimal air pollution and virtually no noise. And it is economical even on a small scale. Several companies are building fuel cells for electric vehicles, for example.[71]

Within two decades, most new buildings could house fuel cells.

Already, Tokyo Electric Power and Southern California Gas have installed fuel cells of between 200 kilowatts and 11,000 kilowatts (11 megawatts) in capacity to provide power and heat in hospitals, hotels, office buildings, and other commercial facilities. Although these systems are small—it would take 500 of them to produce as much power as a standard nuclear plant—the eventual market for fuel cells is likely to demand even tinier ones. A decade or two from now, most new buildings could have natural-gas-powered fuel cells that would not only generate electricity but also replace today's furnace, water heater, and central air conditioner.[72]

Another new generating technology—the photovoltaic (PV) solar cell—shows equally strong potential for decentralization. A photovoltaic system consists of a solid-state, mass-produced panel made of common materials, with no moving parts. Solar PV systems can be deployed on almost any scale, and one of the most compelling applications is on home rooftops. These have already become popular in remote areas lacking access to grid electricity. Rooftop solar systems have been added to more than 200,000 homes in countries such as the Dominican Republic, South Africa, and Sri Lanka. Because more than 2 billion people in developing countries still lack electricity, PVs are likely to play an even larger role in the future. Rooftop photovoltaics also have a role to play in wealthy nations. Norway, for example, already has 50,000 PV-powered country homes.[73]

Government efforts to bring solar-electric buildings into the home market accelerated in the early nineties. Government scientists and entrepreneurs have built several dozen grid-integrated photovoltaic buildings, demonstrating some of the potential of this technology. Germany's Thousand Roofs program was recently upgraded to 2,500 roofs; Switzerland has focused on integrating PVs into the facades of commercial buildings. Also, Japanese and U.S. manufacturers have developed experimental "solar tiles" that could become a common roofing material, while a Swiss company is designing a similar product. In Germany, a major producer of architectural glass has developed a semi-transparent "curtain wall" that provides filtered light as well as electricity to buildings. With the world market growing at a rate of 12 percent annually—and with more than 30 companies now manufacturing solar cells worldwide—these systems are expected to become cost-effective shortly after 2000.[74]

Both solar photovoltaics and fuel cells offer utilities opportunities to improve the overall efficiencies of their systems. It is often expensive to get electricity to remote corners of a power grid, and sometimes growing demand forces a utility to lay new distribution lines, which can require rebuilding transformer stations, tearing up streets and disrupting neighborhoods. According to engineers at the Pacific Gas and Electric Company, it sometimes costs the company twice as much to distribute power as it does to generate it in the first place. In such cases, it might be more economical to pay 10 cents per kilowatt-hour of electricity from small, decentralized generators located near customers than to purchase bulk power for 4 cents per kilowatt-hour and pay another 8 cents to get it to customers. PG&E has used a similar set of calculations to justify installing a 500-kilowatt PV plant at its Kerman substation in the San Joaquin Valley. The solar equipment avoids expensive upgrades for the substation and distribution lines, while reducing environmental costs. Engineers estimate that thousands of megawatts of similar installations could be justified throughout the United States. As a start, a coalition of more than 60 utilities plans to deploy 50 megawatts of solar cells during the next 6 years—much of it in small installations.[75]

Decentralized storage has similar advantages. Such systems would help utilities smooth out the inevitable fluctuations in power demand. Scientists are working on a device called a flywheel, which functions like a mechanical battery. The centrifugal motion of the wheel (spinning inside a sturdy case) contains energy that can be converted to electricity as needed. Although it was invented over 100 years ago, the flywheel only became practical with the development of strong, lightweight composite materials. Modern composites can spin in a vacuum at up to 200,000 revolutions per minute, with the potential to store and release energy at an efficiency of more than 90 percent. Scientists at the Lawrence Livermore Laboratory in California are developing a mass-produced device half the size of a clothes washer that could sit in the basement of a single-family home, storing cheap electricity at night (or solar energy during the day) and releasing it to the grid when demand—and prices—are high.[76]

Japanese and U.S. manufacturers have developed "solar tiles" that could become a common roofing material.

Such technologies ultimately will require modifying the economic conventions of utilities. Today, most utilities recognize only one marginal electricity cost, and weigh all potential power projects on the same scale, paying little attention to the location of generators or customers. Although the PG&E example cited above is an extreme case, on average, one-third of the cost of providing electricity to a residential consumer comes from transmission and distribution. If a utility builds small power plants in a customer's own facility, at least some of the distribution cost is avoided. Thus, a utility that purchases power for 6 cents per kilowatt-hour from a remote coal-fired power plant should be willing to buy it from a customer's rooftop solar system for up to 8 cents if that power is used locally. The German government in effect recognized this principle in a 1991 law that requires utilities to buy power from household-scale systems at 90 percent of the retail electricity price.[77]

The same principle should apply to demand-side management since DSM allows a utility to defer investments not only in power plants but in transmission and distribution facilities as well. U.S. utilities currently spend some $11 billion each year to build and upgrade transmission and distribution—one-third more than they spend on new power generation. (See Figure 6.) Assessing and allocating these costs more accurately could yield large savings—not only in building power plants but in upgrading the transformers, switches, and wires that make up the transmission and distribution systems.[78]

Making these assessments, and integrating the new technologies into the power system, may be simplified by a new generation of telecommunications technologies that could eventually reach most if not all electric power users. Direct, two-way communication between power users and the utility's computers, using existing coaxial (television) cables or fiber optic lines, is now feasible. This could allow each coal plant, fuel cell, flywheel, refrigerator, and air-conditioner to be linked so that the grid operates as a single "smart" system, avoiding overloaded lines and turning generators on and off as needed.[79]

In an experiment being conducted in a subdivision of Little Rock, Arkansas, the Entergy Corporation, a large, regional utility company, has installed household-based computer chips that are linked to home television cables. Using this system, Entergy provides "real time" electricity pricing to customers, who can determine precisely when and at what price they buy

U.S. Private Electric Utility Investments in Generation, and in Transmission and Distribution, 1950-92

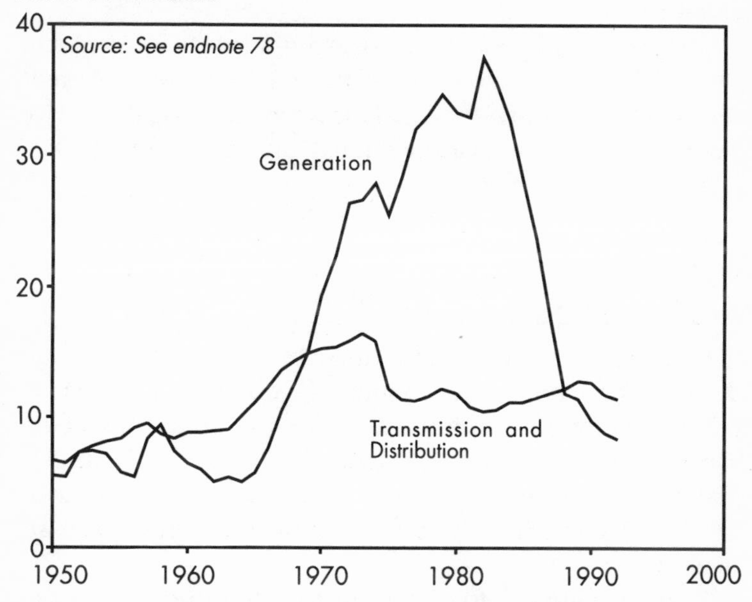

power by programming their air-conditioners, washing machines, and other devices to turn on only when electricity is available at less-expensive rates. Entergy engineers project that this will reduce their peak power demand, avoiding the need to build additional plants. A similar project in Copenhagen is awaiting funding from the European Union.[80]

If deployed properly, small generation and storage devices may increase reliability as well as reduce costs. They also offer developing countries that are plagued by unreliable or incomplete central grids the opportunity to "leapfrog" to twenty-first century power systems, avoiding billions of dollars of investment in central station plants. Carl Weinberg, the former research director for PG&E, who helped develop the concept of a more decentralized power system, observes: "Operating modes for utility systems are likely to evolve along a path similar to that taken by computer networks, telephone switching, and many other large systems....The networks of future utilities will manage many sources, many consumers, and continuous re-evaluation of delivery priorities. All customers and producers will be able to communicate freely through this system to signal changed priorities and costs."[81]

Opportunities presented by demand-side management and decentralized generation suggest the emergence of a different kind of electric utility—one that focuses on meeting the service needs of its customers, largely through investments on the consumer's side of the meter.

The Mirage of Retail Wheeling

For decades, the electric utility industry was one of the world's most staid, immutable businesses, generating little excitement and even less controversy. Today, that industry is being turned upside down as one government after another considers and implements reforms. (See Table 4.) Not since the time of Thomas Edison has the electric power business been so dynamic.[82]

Although utility reforms vary widely, the broader trends are clear: away from public and private monopolies, and toward increased diversity and more competition. The reform efforts reflect a broad consensus that the current system is inefficient and has failed to take advantage of many economic and technological opportunities. Although most such reforms are still at an early stage, they have already yielded clear benefits, including a return to the power industry's historic patterns of falling costs and reliance on less-polluting technologies.

Despite this promising array of innovations, a limited and distorted approach to utility reform has captured the attention of some utility analysts and regulators in the mid-nineties. It goes by the arcane term, "retail wheeling." If enacted in the form proposed in several countries, retail wheeling would only partially realize the benefits of increased competition, and would severely undermine the long-term planning that has been so vital to the evolution of an efficient, environmentally sound electricity market.

At the heart of most retail wheeling proposals is the goal of extending the recognized success of an open market for electricity at the wholesale level all the way to the retail customer, increasing competition and, presumably, lowering costs. Ultimately, all consumers would be encouraged to bargain for their electricity. Since it would not be economical for competing companies to build duplicate distribution lines to serve individual customers, the power produced by an independent producer would have to be "wheeled" across the existing electric utility lines. In arguing their case, retail wheeling proponents point to parallel developments in industries such as rail, telecommunications, and natural gas, each of which has benefitted from increased competition in recent years.[83]

To date, only a few governments have committed to retail wheeling of electricity. The recognized leader is Great Britain, where, in the course of selling its massive, government-owned utility system to private investors, the Thatcher government decided to authorize retail wheeling. At the center of the British reforms was the formation of twelve, government-regulated, regional distribution companies, a national transmission system (jointly owned by the distribution companies), and a semi-competitive generating market. It includes two large private generating companies, one government-owned nuclear generator (whose market share is temporarily protected), and a host of smaller independent power producers.[84]

When a utility sells power to residential consumers, on average one-third of the cost comes from transmission and distribution.

Although the British experiment is still at an early stage, it has already had dramatic and contradictory results: It has succeeded in decimating the British coal industry—which is no longer being protected by subsidized utility purchases—and created a boom in construction of gas-fired power plants. The biggest beneficiaries so far have been the owners of the local dis-

TABLE 4

Electric Utility Reform, Selected Countries

Country	Industry Structure Before Reforms	Description of Reforms
Brazil	Nationally owned, vertically integrated company	Government plans to sell national utility to private investors; independent power producers allowed access to the grid
Chile	Nationally owned, vertically integrated companies	Most of industry already converted to private ownership and vertically separated; independent power production encouraged
Denmark	Private and municipally owned, vertically integrated companies	Utilities required to purchase power from independent wind, biomass, and district heating plants in effort to decentralize generation
Germany	Three dominant, vertically integrated utilities with combined private-public ownership; also smaller municipal companies	Federal government proposing deregulation of power sector; states encouraging small power generation (including wind) and cogeneration; in eastern Germany, municipal utilities competing with large, privately owned western utilities
Hungary	Nationally owned, vertically integrated company	National utility to be vertically separated and sold to private investors; independent power generation will be encouraged
India	Nationally owned generating companies and state-owned generation and distribution companies	Independent power generation, including cogeneration, encouraged; aggressive efforts to promote renewable power sources
Indonesia	Nationally owned, vertically integrated company	Independent power generation encouraged, including biomass and solar power plants
Italy	Nationally owned, vertically integrated company	Utility scheduled to be sold to investors in 1995. Government encouraging cogeneration and renewable power sources

TABLE 4, CONTINUED

Netherlands	Mainly state and municipally owned utilities, with limited vertical integration	Generation and transmission separated from distribution companies; independent generation, wind power encouraged; independent cogeneration booming
New Zealand	National utility for generation and transmission; municipally owned distribution companies	Generation utility separated from transmission company, and from privately and municpally owned distribution companies; retail wheeling allowed for small customers in 1993, large customers in 1994
Norway	Nationally and municipally owned utilities, with some private generation; municipal distribution companies	Retail wheeling allowed for large consumers; small customers still purchase power from municipal distribution companies
Poland	Nationally owned, vertically integrated company	Company slowly being broken up and sold to private investors; plans for retail wheeling. Distribution companies to perform integrated resource planning
Portugal	Nationally owned, vertically integrated company	Private ownership and vertical breakup of industry in process; independent power production promoted
United Kingdom	Nationally owned generation and transmission company, with regional distribution companies	National utility broken up into three generating companies and one transmission company; twelve distribution companies also sold to investors; retail wheeling permitted for large customers now, smaller ones later; independent generators booming; debate continues over demand-side management and broader efficacy of reforms

Source: Worldwatch Institute, based on sources cited in endnote 82

tribution companies, whose profits have skyrocketed. The impact on consumer prices has been more ambiguous, with some rising and others falling. So far, there has been virtually no investment in demand-side management or decentralized generation under the new British system. The ensuing controversy has set off extensive battles over how to "reform the reforms," and it will be several years at least before the long-term structure of the British power industry emerges.[85]

Elsewhere, Norway has adopted a narrower version of retail wheeling, and other countries are considering parallel moves. In the United States, the push for retail wheeling emerged at the state level following enactment of the federal Energy Policy Act of 1992, which opens the utility market to wholesale power wheeling among utilities. This sparked active debate in several states over how to encourage greater wholesale and retail competition. Discussions heated up in early 1994, when the California Public Utilities Commission proposed a radical overhaul of the state's utility regulatory system. The proposal would permit retail wheeling for large industrial customers beginning in 1996, extending to all customers in 2002. The reaction on the part of various "stakeholders" in California was loud and vociferous, setting off an extended series of public debates that may take years to resolve. Already though, shock waves have reached distant corners of the United States: For nearly two decades, California has set the direction of utility reform nationwide.[86]

The stated goals of retail wheeling proponents are laudable: increasing efficiency and lowering costs by spurring competition. However, there are various means of achieving these goals—many of them now being tried in several countries—and it is not clear that retail wheeling would be a step in the right direction. Indeed, some of the proposals tabled so far could undermine other reforms that are still being implemented.

Much of the push for retail wheeling is coming from electricity-intensive industries—such as aluminum, petrochemicals, and autos—that see a short-term opportunity to reduce their own power bills. This opportunity is particularly obvious in the United States, where neighboring utilities often offer prices that vary by as much as 50-100 percent. Most of this disparity is caused by the nuclear plants built during the eighties, which still cost consumers $25-30 billion more each year than they would have to pay for power from a typical new plant running on gas or coal. Even more tempting, utilities in many regions have excess generating capacity, raising the possibility that some companies could purchase power at the marginal cost of operating an idle plant—often as low as 2 cents per kilowatt-hour.[87]

It is hardly surprising that many companies would like to switch to a less-expensive electricity supplier, and conveniently, most proposals for retail wheeling would free large power users first, allowing them to obtain immediate reductions in their power bills. The question for policy makers, however, is whether this particular approach to competition benefits society as a whole. Most of the short-term advantages ascribed to retail wheeling would come simply from reallocating existing costs away from larger industrial customers.[88]

The problem is that these costs would have to be shouldered by someone else—either other power consumers, mainly residential and commercial, or utility company shareholders, who might be forced to pay retroactively for decisions made a decade or more ago. Opponents of retail wheeling argue that shifting costs onto other power consumers would be unfair and economically destructive. Moreover, large power users are hardly a dominant part of today's economy; most new jobs are created by small industrial and commercial companies whose power bills might well rise as a result of retail wheeling.[89]

California regulators have suggested dealing with these potential inequities by charging all electricity consumers, including those who buy their power from another utility or an independent generator, a transmission fee to cover their share of the state's nuclear plants and other high-cost investments. However, the California Commission also promises to bring prices down for industrial customers, and protect utility shareholders from any losses, a set of assurances that simply does not add up. Somebody is bound to lose a lot of money under such a proposal.[90]

Financial markets recognized these risks soon after the California announcement. Moody's Investors Service began an immediate review of the bond ratings of California's utilities, while stock market investors quickly bid down the share prices of California utilities. At the national level, the Dow Jones Utilities index—which covers a sample of major U.S. utilities—plummeted 27 percent between September 1993 and June 1994, reflecting both the growing fervor over retail wheeling and a rise in U.S. interest rates. (See Figure 7.) (During this same period, the broader Dow Jones Industrials index recorded a 3-percent *increase.*) This decline represents some $70 billion in losses to the book value of the U.S. utility industry.[91]

Although not seriously discussed in the California proposal, the most important result of the retail wheeling debate now underway may be to force U.S. utilities to deal honestly with the high-priced nuclear plants on their systems. At least one utility has asked for permission to accelerate the write-off of those "assets," some of which have a book value measured in the billions of dollars, but would probably be worth less than zero if they were offered for sale in today's market. In fact, Shearson Lehman Brothers estimates that nearly one-quarter of all U.S. nuclear plants may be closed for financial reasons during the next decade.[92]

Retail wheeling is fraught with other complications. It inherently conflicts with long-term planning for the power sector, particularly the integrated resource plans required in a growing number of U.S. states. As described earlier, these plans attempt to minimize costs and risks for everyone by evaluating the full range of energy service options, and investing in an appropriate mix of demand- and supply-side technologies, with a diverse spectrum of fuels. They also encourage competition in the generation of electricity and in demand-side management, via competitive bidding for each.[93]

Most retail wheeling proposals would scrap this careful balancing in favor of a commodity market measured only in cents per kilowatt-hour. Such a system would be a step back to the day when utilities worked to sell more power, rather than focusing on electricity services. This would tend to encourage the construction of power plants that produce more pollution. Although the authors of the California proposal claim that

FIGURE 7

Dow Jones Utilities Stock Average, February 1988–June 1994

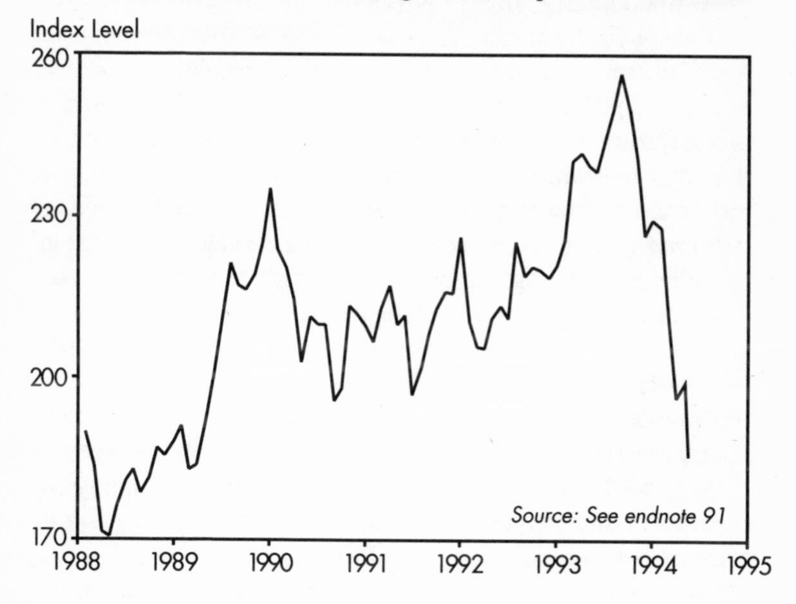

they would somehow compensate for this tendency, it is not clear how they would accomplish it. Moreover, if all such problems were addressed, a utility system based on retail wheeling might turn out to be more complex and confused than those in place today.[94]

As proposed in California, retail wheeling lacks any simple way to incorporate broader societal risks, or even to differentiate the additional value of producing power closer to the customer. The utility would simply be responsible for maintaining the local wires, with little if any planning function, and no incentive to invest in customer-owned options, such as more efficient refrigerators, rooftop solar panels, and the like. Both in Great Britain and California, regulators have promised to encourage utilities to invest in DSM even after retail wheeling is permitted, but so far, they have not proposed any reliable mechanisms to accomplish this. Past efforts to force utilities to invest in efficiency when regulations made it less than profitable for them to do so have been ineffective.[95]

One of the most serious problems with a power system organized around retail wheeling is its incompatibility with environmental goals and policies. As part of the resource planning process, some governments now require that utilities consider the range of pollution costs and other environmental externalities when comparing options for meeting the power needs of customers. But retail wheeling would discourage utilities from investing in anything that costs even slightly more than the least expensive power plant—delaying a range of innovative projects and the commercialization of many new technologies that have only recently gained momentum. This nightmare is already starting to come true: California's retail wheeling proposal has put plans to build 1,350 megawatts of new renewable power projects on hold. In New York, the Niagara Mohawk Company has used the trend to retail wheeling as a reason for reducing its demand-side management spending.[96]

In sum, retail wheeling would probably cause chaos in the power industry for many years, and delay most of the promising reforms now underway. Retail wheeling of the sort proposed in California and Great Britain is unlikely to provide the basis for a new power system that is truly sustainable—economically and politically, as well as environmentally. Still, the intense debate and upheaval unleashed by retail wheeling and other reform proposals may, in the end, contribute to a practical restructuring of the power industry.

Blueprint for Electricity's Future

The question now facing electricity policy makers is whether the advantages of increased competition in the power business can be reconciled with the complex economic and environmental demands of the coming century. We believe that they can. If properly harnessed, the forces of competition have enormous potential to spur efficiency and innovation. However, new models will have to be designed that avoid the pitfalls of slavish devotion either to the "free" market or to *dirigiste* central planning.

A basic blueprint for an efficient, competitive and environmentally responsive power industry can be woven from the many strands of utility reform now found scattered around the world. Additional lessons can be drawn from the restructuring of other industries that share many features with electricity, including telecommunications and natural gas.

Although no one model will work for all nations, we have identified elements that are key to simultaneously meeting economic and environmental needs. These include: a competitive market for wholesale electricity generation; an open-access transmission system; incentives for reliance on diverse power sources, taking into account the environmental differences among them, and development of a service- rather than commodity-oriented local distribution system committed to integrated resource planning and demand-side management.[97]

California's retail wheeling proposal has put plans to build 1,350 megawatts of new renewable power projects on hold.

Already, a consensus has started to gel on the first step in utility reform: open the power generation market to competition, and encourage private generators to enter the arena. As described earlier, this is already permitted in some countries, and has generally delivered reliable and economical electricity. However, no country yet has a fully functioning wholesale power market. Further reforms are needed to assure that electric utilities do not use cross-subsidies to control that market, and to ensure that a diverse mix of companies, offering a range of generating technologies, is able to compete fairly. The move to develop electricity futures markets in Great Britain and the United States is an important step in this direction.

Establishing appropriate "rules of the road" for electricity transmission is one of the most critical elements of an efficient wholesale power market. Since the number of high-voltage power lines that utilities use to move power from generating plants to cities and factories is limited, a competitive power market is only possible if those lines are available to any utility company or independent producer willing to respect the technical limits of the system and pay the prevailing transmission price. The U.S. Energy Policy Act of 1992 is intended to create such a "common carrier" system, but it will be some years before it is fully implemented.[98]

The most critical and controversial question now confronting many countries is how best to organize and regulate the local distribution of power to individual buildings and factories. Long considered the poor stepchild of the electricity business, local distribution is in fact central to any effort to create a truly competitive and sustainable power industry. As a result of the range of innovations now underway in decentralized electricity technologies, power distribution has the potential to become the most dynamic component of the entire business.

Because it makes little economic sense to run multiple power lines down each street—a practice abandoned in the industry's first couple of decades—local electricity distribution remains a natural monopoly. As a result, the utility that owns the local wires must be governed by something other than market forces. In fact, even if retail wheeling is allowed, the distribution process itself would still have to be regulated. But it would be far more efficient and competitive to legally separate the distribution business from power generation and transmission, so as to ensure that local distributors maximize economic and environmental opportunities.

Under our blueprint, local distribution utilities would maintain and integrate local wires, transformers, and other equipment, bargain with independent power producers to purchase wholesale electricity, and help finance demand-side efficiency improvements. The distribution companies would periodically prepare integrated resource plans similar to those produced by some vertically integrated utilities today, selecting an appropriate mix of power service options, including demand-side management and locally-based power generators.

In the United States, a growing number of local power distributors have already discovered the advantages of playing the power market—much the way industrial users would like to. Some small municipal utilities have lowered their costs dramatically by switching suppliers, and other cities are considering forming municipal utilities in order to do the same. (Since they are being arranged by utilities, these transactions are considered wholesale wheeling, which is permitted under *current* U.S. law, though the phenomenon is driven by the same economic disparities that drive retail wheeling.) Although the large price reductions now possible are a function of the current inefficiencies of the power market, they demonstrate the ability of local utilities to bargain just as effectively in competitive wholesale power markets as industrial users might.[99]

The local distribution company is central to creating a competitive and sustainable power industry.

With the right incentives, local distributors can invest in a full range of cost-effective energy services, working closely with energy conservation companies and with sellers of rooftop solar systems and other household generators. In the U.S. Pacific Northwest, for example, where a favorable political climate exists, both public and private utilities have effectively pursued such opportunities. Puget Power and Light, a private utility, and the Emerald People's Utility District, a public one, for example, are investing in everything from more efficient lighting to electronic power controls and rooftop solar systems.[100]

The most comprehensive model of the future distribution utility is found in Sacramento, California. The Sacramento Municipal Utility District (SMUD), which serves a population of several hundred thousand mainly residential and commercial customers, has attracted worldwide attention since 1989, when the concerns of local citizens led to the closing of its only large power plant—the 900-megawatt Rancho Seco nuclear reactor.

SMUD's directors responded to the crisis by hiring a new general manager, S. David Freeman, the maverick former director of the Tennessee Valley Authority. While temporarily relying on wholesale power from distant suppliers, the Sacramento utility is becoming a service-oriented distribution utility.[101]

Among SMUD's innovations: aggressive DSM programs that include replacing 42,000 refrigerators and planting 500,000 shade trees; purchasing power from 4 industrial cogeneration plants; investing in a 50-megawatt wind farm; installing solar-electric systems on customers' rooftops; and investing in electric car development. As a result of such programs, the Sacramento utility has managed to lower the cost of electricity services and transform the once confrontational relationship with its customers. In an industry where executives used to measure success by the size of their latest power plant, SMUD has shown that bigger isn't always better—and that running a local energy service business can be even more rewarding. In the future, utility executives may boast about the number of computer chips installed in customers' homes or the number of solar collectors added to their rooftops.[102]

As demonstrated in Sacramento and other cities, publicly owned utilities are among the pioneers of a new model of local power distribution. Because they are directly responsible to their customers (who are also their owners), getting public utilities to change direction is a straightforward policy decision, uncomplicated by the regulatory formulas needed to guide private utilities. However, there is enormous variation in the caliber and effectiveness of the management of such utilities, the boards of which are sometimes "captured" by particular industries or interest groups that prevent them from serving the broader interests of the community.

Many investor-owned utilities are trying similar innovations, though the process is more complex, requiring an overhaul in regulatory practices. In the United States, for example, regulators have traditionally determined electricity prices on a cost-of-supply basis, allowing utilities to earn profits as a percentage of capital investment. This system is biased toward building power plants and selling more electricity; it provides little incentive for utilities to improve the efficiency of their own operations, let alone their customers' equipment.[103]

New regulatory formulas are clearly required if a different model is to prevail. The key is to decouple a distribution utility's profits from its sales, linking profits instead to the ability to provide energy *services* at the least economic and environmental cost—known as "performance-based" rate-making. It is important to reward good economic and environmental performance, and to reduce utilities' motivation to maximize either capital investments or power sales. Strong, independent government bodies would still need to set and enforce such rules, but market forces would reduce the need for government micromanagement, leading to a more constructive, less adversarial relationship between utilities and their regulators.[104]

Still, new mechanisms will be needed to ensure that planners figure in environmental values as they consider energy service options. If the choice of generating source were to be based on direct costs alone, the next generation of power plants would probably be based entirely on fossil fuels, neglecting renewable energy sources that have higher capital costs today but lower fuel and environmental costs. It is therefore essential that regulators establish a system that counts not just initial investment costs but future environmental ones as well, along with other risk factors—minimizing society's *total* costs by encouraging investment in a variety of generation and end-use technologies. In Massachusetts, the New England Electric System developed a model plan along these lines in 1993, thanks both to regulatory reforms introduced in recent years, as well as the company's own far-sighted leadership.[105]

Two mechanisms have been proposed to address these issues: environmental costing and set-asides. Massachusetts, Nevada, and New York are among the states experimenting with environmental costing. When state regulators evaluate new generating options, they count the cost of pollution against particular projects, which gives renewables an advantage. Taking another approach, California and Great Britain have adopted simple set-asides for renewables, in which a minimum share (10-20 percent) of the market for new power is reserved for "nonfossil" generating technologies. Either one of these approaches can work, and some combination of the two may prove particularly effective. A third, voluntary, approach being tested by a few utilities, including SMUD, is to sell renewably generated electricity at a slightly higher "green" price to consumers who volunteer, allowing them to subsidize development of technologies such as solar cells.[106]

Of course, no single model for electric utility restructuring is likely to work equally well everywhere, and new approaches will have to be adapted to local economic and social conditions. Some countries, for example, may favor private ownership of distribution companies while others will prefer publicly owned systems. The weight placed on environmental costs will also vary widely.

Developing countries and the transitional nations of central Europe face special challenges. The former must provide electricity services for hundreds of millions of people that currently have no power at all; the latter are dismantling electric utility bureaucracies created by communist central planners. In both cases, extensive infrastructure investments and massive institutional changes are needed—offering the opportunity to "leapfrog" to a more efficient power system. Many of these countries have one important advantage: Their old power systems are so discredited that governments are more likely to entertain major reform proposals.

Demand-side management programs make particular sense in developing and central European countries, where electricity is now used so inefficiently. In addition, the decentralized technologies likely to emerge will provide more reliable service in areas now plagued by frequent power outages. Photovoltaic solar cells and other dispersed renewables can play a much larger role in providing electricity to rural dwellers, for example. So far, developing countries have begun to open their power markets to competition, yet no country or foreign aid institution has seriously addressed the critical need to develop a framework for improving energy efficiency and decentralizing the power system.[107]

For the world as a whole, the revolution now underway in the power industry offers many benefits. Like its cousin the telecommunications business, electric power will create large, competitive markets for new, decentralized technologies, and open up a host of business opportunities. In both cases, the structure and regulation of the new industry will take at least a decade to sort out, and the end-result is likely to be far more attractive than the rather messy transition process suggests. The outcome of this transformation will help determine the strength of the global economy and the health of the natural environment for many decades to come.

Notes

1. Electricity as a percentage of final energy demand are Worldwatch Institute estimates, based on Organization for Economic Co-operation and Development (OECD), International Energy Agency (IEA), *Energy Statistics and Balances of Non-OECD Countries 1990-1991* (Paris: 1993), and on D.O. Hall, King's College London, private communication and printout, March 7, 1994; size of the world power industry is a Worldwatch Institute estimate based on United Nations (UN), *1990 Energy Statistics Yearbook* (New York: 1992), and on Edison Electric Institute (EEI), *Statistical Yearbook of the Electric Utility Industry 1991* (Washington, D.C.: 1992); size of world auto industry is a Worldwatch Institute estimate based on Department of Commerce (DOC), Bureau of the Census (BOC), *Statistical Abstract of the United States 1992* (Washington, D.C.: 1992), on American Automobile Manufacturers Association, *World Motor Vehicle Data 1993* (Detroit, Mich.: 1993), and on Motor Vehicles Manufacturers Association of the United States, Inc., *Facts & Figures '90* (Detroit, Mich.: 1990).

2. Worldwatch Institute estimates, based on OECD, IEA, op. cit., note 1; Worldwatch Institute estimate, based on Jane Dignon and Sultan Hameed, "Global Emissions of Nitrogen and Sulfur Oxides from 1860 to 1980," *APCA Journal*, February 1989, on Dr. J. Dignon, Lawrence Livermore National Laboratory, Livermore, Calif., private communication and printout, February 23, 1994, on British Petroleum (BP), *BP Statistical Review of World Energy* (London: 1993), and on U.S. Environmental Protection Agency, *National Air Pollutant Emissions Trends, 1900-1992* (Research Triangle Park, N.C., 1993).

3. Linda Bromley, Energy Information Administration (EIA), U.S. Department of Energy (DOE), Washington, D.C., private communication, December 13, 1993.

4. Richard Munson, *The Power Makers* (Emmaus, Pa.: Rodale Press, 1985).

5. Richard F. Hirsh, *Technology and Transformation in the American Electric Utility Industry* (New York: Cambridge University Press, 1989); Munson, op. cit., note 4.

6. Munson, op. cit., note 4.

7. Colin Ashmore, "Policies, Politics, and Prices," *Electricity International*, April 1994.

8. Electricity costs are from Hirsh, op. cit., note 5, from EEI, "EEI Pocketbook of Electric Utility Industry Statistics," Washington, D.C., 1992, and from EEI, *Historical Statistics of the Electric Utility Industry Through 1970* (New York: 1973).

9. Efficiency of steam turbine plants is from Hirsh, op. cit., note 5, and from *Electric Light and Power*, various issues.

10. Irvin C. Bupp and Jean-Claude Derian, *Light Water: How the Nuclear Dream Dissolved* (New York: Basic Books, Inc., 1978); International Atomic Energy Agency, *Nuclear Power Reactors in the World* (Vienna: 1993); Greenpeace International, WISE-Paris, and Worldwatch Institute, *World Nuclear Industry Status Report: 1992* (London: Greenpeace International, 1992).

11. Figure 1 is based on UN, *World Energy Supplies 1950-1974* (New York: 1976), on OECD, IEA, *Energy Balances of OECD Countries* (Paris: various years), on International Monetary Fund, *World Economic Outlook May 1993* (Washington, D.C.: 1993), on Robert Summers and Alan Heston, "The Penn World Table (Mark 5): An Expanded Set of International Comparisons, 1950-1988," *Quarterly Journal of Economics*, May 1991 (based on purchasing power parity), and on Yoko Takahashi, IEA, Paris, private communication, April 29, 1994.

12. French debt from "Export Surge Propels EdF from Strength to Strength," *European Energy Report*, February 18, 1994; Richard Rudolph and Scott Ridley, *Power Struggle: The Hundred-Year War Over Electricity* (New York: Harper & Row, 1986).

13. Forty-five percent from Public Service Company of New Mexico, "San Juan Generating Station: Fact Sheet," February 20, 1992, in Charles Bensinger, "Solar Thermal Repowering: A Technical and Economic Pre-Feasibility Study" (draft revised version 2.0), The Energy Foundation, San Francisco, Calif., 1993; Barbara J. Cummings, *Dam the Rivers, Damn the People* (London: Earthscan Publications Ltd., 1990).

14. World Bank, *The World Bank's Role in the Electric Power Sector* (Washington, D.C.: 1993); Stefan Wagstyl, "Power Cuts 'Crippling India's Prospects,'" *Financial Times*, October 29, 1993.

15. World Bank, op. cit., note 14.

16. "Germany Decides on DM7bn/yr Coal Aid Plus Kohlepfennig Rise," *European Energy Report*, November 12, 1993; Jennifer S. Gitlitz, "The Relationship Between Primary Aluminum Production and the Damming of World Rivers," Energy and Resources Group, University of California, Berkeley, Calif., July 21, 1993; François Nectoux, *Crisis in the French Nuclear Industry* (Amsterdam: Greenpeace International, 1991).

17. Policy Topic Committee of the Association of Demand Side Management Professionals, "Status Report: State Requirements for Considering Environmental Externalities in Electric Utility Decision-Making," Boca Raton, Fla., 1993.

18. Carbon dioxide emission percentage is a Worldwatch Institute estimate, based on OECD, IEA, op. cit., note 1; United Nations, Report of the Intergovernmental Negotiating Committee for a Framework Convention on Climate Change, Fifth Session, Second Part, New York, April 30-May 9, 1992; Interim Secretariat of the United Nations Framework Convention on Climate Change, "Status of Ratification of the United Nations Framework Convention on Climate Change" (electronic bulletin board posting), April 15, 1994.

19. Christopher Flavin, *Electricity's Future: The Shift to Efficiency and Small-Scale Power*, Worldwatch Paper 61 (Washington, D.C.: Worldwatch Institute, November 1984).

20. EEI, *Capacity and Generation of Non-Utility Sources of Energy* (Washington, D.C.: various years). Figure 2 is based on DOE, EIA, *Annual Electric Generator Report* (electronic database) (Washington, D.C.: 1993), and on EEI, op. cit., this note; independent producer data through 1992 is the amount of capacity placed in operation each year and still operating in 1992; thus the data underestimate the total amount of new capacity that was installed annually by independent producers; U.S. Congress, "National Energy Policy Act of 1991," Public Law 102-486, Washington, D.C., October 24, 1992; Mary O'Driscoll, "Nymex Executive Outlines Draft Electricity Futures Contract," *The Energy Daily*, April 22, 1994.

21. DOE, EIA, op. cit., note 20; EEI, op. cit., note 20.

22. U.S. Agency for International Development (USAID), Office of Energy and Infrastructure (OEI), "Country Profiles on India, Indonesia, and Pakistan," *Private Power Reporter*, March 1993; "Netherlands," Country Profiles, *European Energy Report*, March 1992; "The Power and the Glory," *The Guardian*, February 1, 1992; Ministry of Energy, Danish Energy Agency, *Energy Efficiency in Denmark* (Copenhagen: 1992); Sara Knight, "German Survey," *Windpower Monthly*, March 1993; "Germany," Country Profiles, *European Energy Report*, May 1992; Masao Karube, New Energy Foundation, Tokyo, private communication, June 6, 1994; Steven Strong, "An Overview of Worldwide Development Activity in Photovoltaics," Solar Design Associates, Harvard, Mass., 1993; Andrew Holmes, "Electricity in Europe: Power and Profit," *Financial Times Management Report*, London, 1990; Sandy Hendry, "US, Europe Firms Look to Spark China Power Ventures," *Journal of Commerce*, June 2, 1993; Andy Pasztor, "Power Plants in Mexico Cast Pall Over Nafta," *Wall Street Journal*, September 9, 1993; "News Briefs," *Electricity International*, February 1994.

23. Sir Leon Brittan, "Competition in the Electricity and Gas Markets," *Target 1992*, June 1991; "Brussels Sues Six Member States in Drive for EU Deregulation," *European Energy Report*, March 1994; "Denmark Ends Gas Monopoly Following Brussels Intervention," *European Energy Report*, April 29, 1994; "New German Policy Born Amid Coal Concerns and Nuclear Controversy," *European Energy Report*, December 10, 1993.

24. Value of process steam from cogeneration plants is a Worldwatch Institute estimate based on natural gas prices from DOE, EIA, *Monthly Energy Review September 1993* (Washington, D.C.: U.S. Government Printing Office (GPO), 1993).

25. Source for current industrial cogeneration in U.S. is "Manufacturing Energy Consumption Survey: Preliminary Estimates, 1991," in DOE, EIA, op. cit., note 24; EEI, op. cit., note 21; Neil Slavin, Intelligen Energy Systems, Westford, Mass., private communication, February 14, 1994; Preben Maegaard, Folkecenter for Renewable Energy, Hurup Thy, Denmark, private communication, April 21, 1994.

26. Lester P. Silverman, McKinsey & Co., Inc., Washington, D.C., private communication, October 13, 1993.

27. Robert H. Williams and Eric D. Larson, "Expanding Roles for Gas Turbines in Power Generation," in Thomas B. Johansson, Birgit Bodlund, and Robert H. Williams, eds., *Electricity: Efficient End-Use and New Generation Technologies, and Their Planning Implications* (Lund, Sweden: Lund University Press, 1989); Robert L. Bradley, Jr., "Reconsidering the Natural Gas Act," Southern Regulatory Policy Institute Issue Paper No. 5, Roswell, Ga., August 1991; "The Use of Natural Gas in Power Stations," *Energy in Europe* (Commission of the European Communities, Brussels), December 1990.

28. Figure 3 uses "best U.S. steam" data from Hirsh, op. cit., note 5, and from *Electric Light and Power*, various issues, bases "average" data on DOC, BOC, *Historical Statistics of the United States: Colonial Times to 1970*, Part 2 (Washington, D.C.: 1975), on DOE, EIA, *Annual Energy Review 1992* (Washington, D.C.: GPO, 1993), and on DOE, EIA, *Monthly Energy Review April 1994* (Washington, D.C.: GPO, 1994), and bases combined-cycle gas turbine data on private communications with various sources, including Bill Brooks, Virginia Power, Chester, Va., October 8, 1993, and Bob Bjorge, General Electric, Schenectady, N.Y., private communication and printout, August 26, 1993; conversion efficiency levels for all plants are determined using the higher heating value, which gives lower efficiency levels; David S. Bazel, ABB, North Brunswick, N.J., private communication and printout, November 2, 1993.

29. Eric Jeffs, "First 9F in Service with EdF," *Electricity International*, June/July 1993; Steven Collins, "Special Report: Gas Fired Powerplants," *Power Magazine*, February 1993; "For the Record," *Energy Economist*, April 1993; "For the Record," *Energy Economist*, September 1993; Neil Buckley, "Hurdles in the Path of the Dash for Gas," *Financial Times*, December 10, 1992; Teeside configuration is from Kristin Rankin, Enron Corp, Houston, Tex., private communication, October 20, 1993.

30. Williams and Larson, op. cit., note 27; Bjorge, op. cit., note 28; William H. Day and Ashok D. Rao, "FT4000 HAT with Natural Gas Fuel," *Turbomachinery International*, January/February 1993; Jack Janes, California Energy Commission, Sacramento, Calif., private communication, May 31, 1994; Steven Collins, "Small Gas Turbines Post Gains in Performance," *Power Magazine*, October 1992.

31. Table 2 is based on Bjorge, op. cit., note 28, on M.W. Horner, "GE Aeroderivative Gas Turbines—Design and Operating Features," GE Aircraft Engines, GE Power Generation, Evendale, Ohio, 1993, on John C. Trocciola, International Fuel Cells, South Windsor, Conn., private communication, January 3, 1994, and on Bazel, op. cit., note 28. The first two coal plants are using pulverized coal, and the fuel cell plant is a phosphoric acid fuel cell using hydrogen reformed from natural gas; conversion efficiency levels for all plants are determined using the higher heating value, which gives lower efficiency levels; D.L. Chase, J.M. Kovacik, and H.G. Stoll, "The Economics of Repowering Steam Power Plants," General Electric Company, 1992; Pamela Newman, "TVA Considers Natural Gas at Nuclear Plants," *The Energy Daily*, April 20, 1994; Gero Lücking et al., "Essential Elements in the Ecological Reform of the Energy Industry in Ukraine," Öko Institut, Berlin, March 1994.

32. Douglas M. Todd and Robert M. Jones, General Electric, "Advanced Combined Cycles Provide Economic Balance for Improved Environmental Performance," presented at International Power Generation Conference, San Diego, Calif., October 6-10, 1991; William Keeling, "Indonesia's Power Scramble," *Financial Times*, August 10, 1993; GE Power Generation, "MS9001E Gas Turbines: Heavy-Duty 50 Hz Power Plant," Schenectady, N.Y., 1991; "Europe's Most Modern Combined-Cycle Plant," *Electricity International*, June/July 1993.

33. Jeffs, op. cit., note 29; D.M. Todd, "Clean Coal Technologies for Gas Turbines," GE Power Generation, Schenectady, N.Y., July 1993; Bjorge, op. cit., note 28.

34. Ragnar Lundqvist, Martti Puhakka, and Krister Ståhl, "Pressurized CFB Biomass Gasification—The Bioflow Energy System," presented to the CFB4 Conference, Somerstet, Penn., August 1-5, 1993; Ragnar Lundqvist, Bioflow/Ahlstrom, Varkaus, Finland, private communication, January 27, 1994; Ragnar Lundqvist, "The IGCC Demonstration Plant at Värnamo," *Bioresource Technology*, Vol. 46, pp. 49-53, 1993; A.E. Carpentieri, E.D. Larson, and J. Woods, "Prospects for Sustainable, Utility-scale, Biomass-based Electricity Supply in Northeast Brazil," Center for Energy and Environmental Studies, Princeton University, July 1992; Philip Elliott and Roger Booth, "Brazilian Biomass Power Demonstration Project," Special Project Brief, Shell International, London, July 1993.

35. Carbon dioxide emissions are essentially zero if new biomass is grown to replace that burned; Walt Patterson, *Power from Plants: The Global Implications of New Technologies for Electricity from Biomass* (London: Royal Institute of International Affairs and Earthscan Publications Ltd., 1994); 30 percent of electricity is a Worldwatch Institute estimate based on D.O. Hall et al., "Biomass for Energy: Supply Prospects," in Thomas B. Johansson et al., eds., *Renewable Energy: Sources for Fuels and Electricity* (Washington, D.C.: Island Press, 1993), Population Reference Bureau (PRB), "World Population Estimates and Projection by Single Years: 1750-2100," Washington, D.C., 1992, and UN, op. cit., note 1.

36. Janice G. Hamrin and Nancy Rader, "Non-Utility Power Development in the USA," *Energy Policy*, November 1992; California renewables capacity is from Jan Hamrin and Nancy Rader, *Investing in the Future: A Regulator's Guide to Renewables* (Washington, D.C.: National Association of Regulatory Utility Commissioners (NARUC), 1993); percent of electricity supply is from Karen Griffin, California Energy Commission (CEC), private communication and printout, April 26, 1994.

37. Ronald DiPippo, "Geothermal Energy," *Energy Policy*, October 1991; World geothermal capacity is a Worldwatch estimate based on UN, *1991 Energy Statistics Yearbook* (New York: 1993), and on Gerald Huttrer, Geothermal Management Company, Inc., Frisco, Colo., private communication, February 10, 1994; Civis G. Palmerini, "Geothermal Energy," in Johansson et al., op. cit., note 35.

38. DOE, "U.S. Geothermal Energy R&D Program Multi-Year Plan, 1988-1992," Washington, D.C., 1988; Edwin Karmiol, "Japan Awakening to Potential of Domestic Geothermal Energy," *The Solar Letter*, November 13, 1992.

39. Birger Madsen, BTM Consult ApS, Ringkoeping, Denmark, private communication to René Karottki, Forum for Energy and Development, Copenhagen, March 21, 1994; Paul Gipe, Paul Gipe and Associates, Tehachapi, Calif., private communication and printout, April 6, 1994; Christopher Flavin and Nicholas Lenssen, *Power Surge: Guide to the Coming Energy Revolution* (New York: W.W. Norton & Company, in press).

40. Susan Hock, Robert Thresher, and Tom Williams, "The Future of Utility-Scale Wind Power," in S. Burley and M.E. Arden, eds., *Advances in Solar Energy: An Annual Review of Research and Development* (Boulder, Colo.: American Solar Energy Society, 1992).

41. D.L. Elliott, L.L. Windell, and G.L. Gower, *An Assessment of the Available Windy Land Area and Wind Energy Potential in the Contiguous United States* (Richland, Wash.: Pacific Northwest Laboratory (PNL), 1991); William Babbitt,

Associated Appraisers, Cheyenne, Wyo., private communication, October 11, 1990; Paul Gipe, "Wind Energy Comes of Age," Paul Gipe and Associates, Tehachapi, Calif., May 13, 1990.

42. Michael J. Grubb and Niels I. Meyer, "Wind Energy: Resources, Systems and Regional Strategies," in Johansson et al., op. cit., note 35; Elliott, Windell, and Gower, op. cit., note 41; land use figures are for Class 3 wind areas and above, which includes area with a wind power density at a height of 50 meters of 300-400 watts per square meter and an average annual wind speed of at least 6.4 meters per second (14 miles per hour).

43. Flavin and Lenssen, op. cit., note 39; Grubb and Meyer, op. cit., note 42.

44. Pat De Laquil III et al., "Solar-Thermal Electric Technology," in Johansson et al., op. cit., note 35; David Kearney, KJC Operating Company, Kramer Junction, Calif., private communication and printout, July 30, 1993; number of homes is Worldwatch Institute estimate based on Don Logan, Luz International Limited, Los Angeles, Calif., private communication, September 26, 1990, and DOC, BOC, *Statistical Abstract of the United States 1990* (Washington, D.C.: U.S. GPO, 1990); Michael Lotker, "Barriers to Commercialization of Large-Scale Solar Electricity: Lessons from the LUZ Experience," Sandia National Laboratories, November 1991; Gabi Kennen, Solel, Jerusalem, private communication and printout, June 24, 1993; David Mills, University of Sydney, private communication and printout, April 5, 1994.

45. Electricity cost from parabolic troughs has been converted to 1993 dollars and is from DOE, Solar Thermal and Biomass Power Division, "Solar Thermal Electric Technology Rationale," Washington, D.C., August 1990; Idaho National Engineering Laboratory et al., *The Potential of Renewable Energy: An Interlaboratory White Paper*, prepared for the Office of Policy, Planning and Analysis, DOE, in support of the National Energy Strategy (Golden, Colo.: Solar Energy Research Institute, 1990); Stephen Kaneff, "Mass Utilization of Solar Thermal Energy," Energy Research Centre, Australian National University, Canberra, September 1992; Mills, op. cit., note 44.

46. De Laquil III et al., op. cit., note 44.

47. Solar resource of 2.5 million exajoules per year is from Denis Hayes, *Rays of Hope: The Transition to a Post-Petroleum World* (New York: W.W. Norton & Company, 1977), and is total resource before using a 15 percent conversion efficiency for sunlight to electricity; electricity consumption is from OECD, IEA, op. cit., note 1; solar thermal power land use is from David Mills, University of Sydney, private communication and printout, March 25, 1994; electricity generation is from OECD, IEA, op. cit., note 1, and DOE, EIA, *Monthly Energy Review*, April 1994, op. cit., note 28; U.S. land area used by the military from Michael Renner, "Assessing the Military's War on the Environment," in Lester R. Brown et al., *State of the World 1991* (New York: W.W. Norton & Company, 1991).

48. Henry Kelly and Carl J. Weinberg, "Utility Strategies for Using Renewables," in Johansson et al., op. cit., note 35.

49. Lawrence Flowers et al., "Utility-Scale Wind Energy Update," in Burley and Arden, op. cit., note 40; M.N. Schwartz, D.L. Elliott, and G.L. Gower, PNL,

"Seasonal Variability of Wind Electric Potential in the United States," presented at the American Wind Energy Association's Windpower '93 Conference, San Francisco, Calif., July 12-16, 1993; Don Smith and Mary Ilyin, Pacific Gas and Electric, "Wind and Solar Energy: Costs and Value," *Proceedings of ASME 10th Wind Energy Symposium*, Houston, January 1991.

50. Grubb and Meyer, op. cit., note 42; Kelly and Weinberg, op. cit., note 48; Carl Weinberg, Weinberg and Associates, Walnut Creek, Calif., private communication, April 13, 1994.

51. Christopher Hocker, "The Miniboom in Pumped Storage," *Independent Energy*, March 1990; Leslie Lamarre, "Alabama Cooperative Generates Power from Air," *EPRI Journal*, December 1991; David Mills and Bill Keepin, "Baseload Solar Power: Near-term Prospects for Load Following Solar Thermal Electricity," *Energy Policy*, August 1993.

52. Kelly and Weinberg, op. cit., note 48; David Mills, University of Sydney, private communication and printout, April 5, 1994.

53. Table 3 is based on DOE, EIA, *Historical Plant Cost and Annual Production Expenses for Selected Electric Plants 1987* (Washington, D.C.: GPO, 1989), on DOE, EIA, *Electric Plant Cost and Power Production Expenses* (Washington, D.C.: GPO, various years), on Charles Komanoff, Komanoff Energy Associates, New York, private communications and printout, February 9, 1989 and June 9, 1993, and on Flavin and Lenssen, op. cit., note 39.

54. "BRPU Competition Lowers Cost of Electricity," *Coalition Energy News*, Winter 1994; in April 1994, the California Public Utilities Commission, announced that the winning bids would be put on hold while the Commission considered regulatory reforms.

55. David Roe, *Dynamos and Virgins* (New York: Random House, 1984); Hirsh, op. cit., note 5.

56. OECD, IEA, *Energy Policies of IEA Countries, 1991 Review* (Paris: 1992).

57. Amory B. Lovins, *Soft Energy Paths: Toward a Durable Peace* (Cambridge, Mass.: Ballinger Publishing Company, 1977); OECD, IEA, op. cit., note 1; Arnold P. Fickett, Clark W. Gellings, and Amory B. Lovins, "Efficient Use of Electricity," *Scientific American*, September 1990; Steven Nadel, Virendra Kothari, and S. Gopinath, "Opportunities for Improving End-Use Electricity Efficiency in India," American Council for an Energy-Efficient Economy (ACEEE), Washington, D.C., November 1991; U.S. Congress, Office of Technology Assessment (OTA), *Energy Efficiency Technologies for Central and Eastern Europe* (Washington, D.C.: GPO, 1993).

58. Chris J. Calwell and Ralph C. Cavanagh, "The Decline of Conservation at California Utilities: Causes, Costs and Remedies," Natural Resources Defense Council (NRDC), San Francisco, Calif., July 1989; David Moskovitz, "Profits and Progress Through Least-Cost Planning," NARUC, Washington, D.C., November 1989.

59. Rowe quote is from EEI, *Washington Letter*, Washington, D.C., September 15, 1989; Moskovitz, op. cit., note 58; Stephen Wiel, "Making Utility Efficiency

Profitable," *Public Utilities Fortnightly*, July 1989; Michael Smith, "An Island's Experience," *Financial Times*, September 8, 1993.

60. Cynthia Mitchell, consulting economist, Reno, Nev., private communication, May 13, 1994; Cynthia Mitchell, "Integrated Resource Planning Survey: Where the States Stand," *The Electricity Journal*, May 1992; Electric Power Research Institute (EPRI), "Amorphous Metal Transformers Cited as Way to Improve Energy Efficiency," Palo Alto, Calif., news release, April 28, 1993; Allied-Signal Inc., "Improving America's Energy Distribution System," Washington, D.C., undated.

61. Figure 4 is based on DOE, EIA, *Electric Power Annual 1991* (Washington, D.C.: GPO, 1993), and on Bromley, op. cit., note 3; California from Ralph Cavanagh, "The Great 'Retail Wheeling' Illusion—and More Productive Energy Futures," E-SOURCE, Boulder, Colo., 1994, which cites document from the California Public Utilities Commission; EPRI estimates from Clark Gellings, EPRI, as cited in Amory Lovins, "Apples, Oranges, and Horned Toads: Is the Joskow & Marron Critique of Electric Efficiency Costs Valid?" *The Electricity Journal*, May 1994; Paul L. Joskow and Donald B. Marron, "What Does Utility-Subsidized Energy Efficiency Really Cost?" *Science*, April 16, 1993; Amory B. Lovins, "The Cost of Energy Efficiency" (letter to the editor), *Science*, August 20, 1993.

62. John Fox, Ontario Hydro, Toronto, Ont., Canada, private communication, September 29, 1993; Katrina van Bylandt, Power Smart, Inc., Vancouver, B.C., private communication, January 28, 1993; Alliance to Save Energy, "Quebec Utility Launches One of North America's Largest Residential Energy Conservation Programs," press release, Washington, D.C., December 13, 1993; Evan Mills, "Efficient Lighting Programs in Europe: Cost Effectiveness, Consumer Response, and Market Dynamics," *Energy—The International Journal*, Vol. 18, No. 2, 1993.

63. Evan Mills, Lawrence Berkeley Laboratory, Berkeley, Calif., private communication, June 15, 1993; Mills, op. cit., note 62; Wim Sliepenbeek, "Massive Programs Get the Dutch Market Moving," *IAEEL Newsletter* (International Association for Energy-Efficient Lighting, Stockholm), No. 1, 1993; Uwe Leprich, "German Giant Explores Its Demand-Side Resources," *IAEEL Newsletter*, No. 2, 1992; Reinhard Loske, Wuppertal Institute for Climate, Environment and Energy, Wuppertal, Germany, private communication, September 24, 1993; Uwe Fritsche, Öko Institut, Darmstadt, Germany, private communication, May 17, 1994.

64. Thailand from Peter du Pont, Terry Kraft-Oliver, and Peter Rumsey, International Institute for Energy Conservation (IIEC), Washington, D.C., private communication, June 23, 1993; Howard S. Geller and José Roberto Moreira, "Brazil Encourages Electricity Savings," *Forum for Applied Research and Public Policy*, University of Tennessee, Fall 1993; Mark D. Levine, Feng Liu, and Jonathan E. Sinton, "China's Energy System: Historical Evolution, Current Issues, and Prospects," in *Annual Review of Energy and the Environment 1992* (Palo Alto, Calif.: Annual Reviews Inc., 1992); Ignacio Rodriguez and David Wolcott, "Growth Through Conservation: DSM in Mexico," *Public Utilities Fortnightly*, August 1, 1993; Lawrence Berkeley Laboratory estimate is from Mark D. Levine et al., *Energy Efficiency, Developing Nations, and Eastern Europe*, A Report to the U.S. Working Group on Global Energy Efficiency (Washington, D.C.: IIEC, 1991), and from Charles Campbell, Lawrence Berkeley Laboratory, Berkeley, Calif, private communication and printout, June 19, 1992; Howard Geller, *Efficient Electricity Use: A Development Strategy for Brazil* (Washington, D.C.: ACEEE, 1991); Lee Schipper and Eric Martinot, "Decline and Rebirth: Energy Demand in the Former Soviet Union" (draft), Paper II: Towards Efficiency in 2010, Lawrence Berkeley Laboratory, Berkeley, Calif., September 1992; David Wolcott, Jaroslaw Dybowski, and Ewaryst Hille, "Implementing Demand-Side Management Through Integrated Resource Planning in Poland," presented at the European Council for an Energy-Efficiency Economy Summer Study, Rungstedgaard, Denmark, May 1993.

65. Figure 5 is based on DOC, BOC, *Statistical Abstract of the United States* (Washington, D.C.: various years), on DOC, *Department of Commerce News*, March 14, 1994, on DOE, EIA, *Annual Energy Review 1992*, op. cit., note 28, on DOE, EIA, *Monthly Energy Review April 1994*, op. cit., note 28, on DOE, EIA, *State Energy Data Report: Consumption Estimates* (Washington, D.C.: GPO, various years), on PRB, *World Population Data Sheet* (Washington, D.C.: various years), and on Andrea Gough, CEC, Sacramento, Calif., private communication and printout, May 17, 1994; "Holland Turns the Tide," *IAEEL Newsletter*, No. 1, 1992; EPRI, *Drivers of Electricity Growth and the Role of Utility Demand-Side Management* (Palo Alto: 1993); Greg M. Rueger, senior vice president and general manager, Pacific Gas and Electric Company, Testimony before the Subcommittee on Energy and Power, Committee on Commerce and Energy, U.S. House of Representatives, Washington, D.C., March 7, 1991; Eric Hirst, "Managing Demand for Electricity: Will It Pay Off?" *Forum for Applied Research and Public Policy*, Fall 1992.

66. Matthew L. Wald, "Utilities Offer $30 Million for a Better Refrigerator," *New York Times*, July 8, 1993; Michael L'Ecuyer et al., "Stalking the Golden Carrot: A Utility Consortium to Accelerate the Introduction of Super- Efficient, CFC-Free Refrigerators," in *ACEEE 1992 Summer Study on Energy Efficiency in Buildings* (Washington, D.C.: ACEEE, 1992); John Feist, Super Efficient Refrigerator Program, Washington, D.C., private communication, April 11, 1994.

67. Michael Philips, *The Least Cost Energy Path for Developing Countries: Energy Efficient Investments for the Multilateral Development Banks* (Washington, D.C.: IIEC, 1991); Environmental Defense Fund (EDF) and NRDC, *Power Failure: A Review of the World Bank's Implementation of its New Energy Policy* (New York: EDF, 1994); Kim Coghill, "Bank to Factor in Environmental Costs When Assessing Asian Power Projects," *Journal of Commerce*, September 22, 1993.

68. DOE, EIA, *Annual Energy Review 1992*, op. cit., note 28.

69. Worldwatch estimate, assuming a fuel efficiency equivalent to the GM Impact's of .35 kilowatt-hours per kilometer for electrics, and based on DOE, EIA, *Annual Energy Review 1992*, op. cit., note 28, and on Department of Transportation, Federal Highway Administration, *Highway Statistics 1991* (Washington, D.C.: 1992); Diane Wittenberg, Southern California Edison, Rosemead, Calif., private communication, June 1, 1994.

70. Brooke Stoddard, "Fuel Cell Update," *American Gas*, June 1993.

71. Fuel Cell Commercialization Group, "What Is a Fuel Cell?" Washington, D.C., 1992; Philip H. Abelson, "Applications of Fuel Cells" (editorial), *Science*, June 22, 1990; Robert H. Williams, "The Clean Machine," *Technology Review*, April 1994.

72. John Douglas, "Utility Fuel Cells in Japan," *EPRI Journal*, September 1991; "SoCalGas Inks First Commercial Fuel Cell Deal," *Oil & Gas Journal*, April 15, 1991.

73. Neville Williams, Solar Electric Light Fund (SELF), Washington, D.C., private communication, January 14, 1994; Derek Lovejoy, "Electrification of Rural Areas by Solar PV," *Natural Resources Forum*, May 1992; Mark Hankins, *Solar Rural Electrification in the Developing World* (Washington, D.C.: SELF, 1993); Strong, op. cit., note 22.

74. Strong, op. cit., note 22; Ronal W. Larson, Frank Vignola, and Ron West, "Economics of Solar Energy Technologies," American Solar Energy Society, Boulder, Colo., December 1992; PV market growth is an average for the past five years, and from Paul Maycock, "1993 World Module Shipments," *Photovoltaic News*, February 1994.

75. Tammie R. Candelario and Tim Townsend, "PVUSA—Progress and Plans," in Burley and Arden, op. cit., note 40; "Siemens Completes 500-Kilowatt (Peak) PV System at Kerman, Calif., for PG&E," *The Solar Letter*, May 14, 1993; Carl Weinberg, Joseph J. Iannucci, and Melissa M. Reading, "The Distributed Utility: Technology, Customer and Public Policy Changes Shaping the Electrical Utility of Tomorrow," PG&E Research and Development, San Ramon, Calif., December 1992; Utility PhotoVoltaic Group, "Electric Utilities Serving 40% of U.S. Consumers Propose $513 Million Program to Accelerate Use of Solar Photovoltaics," Washington, D.C., September 27, 1993.

76. Richard F. Post, "Flywheel Energy Storage," *Scientific American*, September 1973; John V. Coyner, "Flywheel Energy Storage and Power Electronics Program," Oak Ridge National Laboratory, 1993; Abacus Technology Corporation, *Technology Assessments of Advanced Energy Storage Systems for Electric and Hybrid Vehicles*, prepared for DOE, April 30, 1993; Lawrence Livermore projections from Richard F. Post, Lawrence Livermore National Laboratory, Livermore, Calif., private communication, July 15, 1993.

77. Weinberg, Iannucci, and Reading, op. cit., note 75; Germany from Sara Knight, "Portrait of a Booming Market," *Windpower Monthly*, March 1993.

78. Figure 6 is based on EEI, "Pocketbook of Utility Statistics," op. cit., note 8, and on EEI, *Statistical Yearbook of the Electric Utility Industry* (Washington, D.C.: various years).

79. Steven R. Rivkin, "Look Who's Wiring the Home Now," *New York Times Magazine*, September 26, 1993.

80. "Advanced Electronic Energy Consumer Management System for Renovated Housing Block with District Heating Supply," proposal to the Commission of the European Communities, Director-General for Energy, THERMIE, Brussels, November 1992.

81. Carl J. Weinberg and Katie McCormack, "Toward a Sustainable Energy Future," speech to Consumer Federation of America, Washington, D.C., May 27, 1993; Leslie Lamarre, "The Vision of Distributed Generation," *EPRI Journal*, April/May 1993.

82. Table 4 is based on the following sources: "Brazil to Privatize Power Firms, Report Says," *Journal of Commerce*, November 9, 1993; USAID, OEI, "Focus on South America," *Private Power Reporter*, Washington, D.C., December 1993; World Bank, op. cit., note 14; "Denmark Plans for the 1990s," *Power in Europe*, April 12, 1990; "Mini-boom with Mini-CHP Plant," *European Energy Report*, March 4, 1994; Judy Dempsey, "Germany Wrestles with Energy Puzzle," *Financial Times*, May 12, 1994; "Hungary Passes Privatization Bill; 1,530 MW to be Offered by Year-End," *Electric Utility Week*, April 25, 1994; "Electricity," *Energy Economist*, December 1993; USAID, OEI, op. cit., note 22; "Italy," *European Energy Report*, Supplement, April 1994; Jan Paul van Soest, Center for Energy Conservation and Environmental Technologies, Delft, private communication, May 17, 1994; Molly Melhiush, Energy Watch, York Bay Eastborne, New Zealand, private communication, May 16, 1994; David Wolcott, RCG/Hagler, Bailly, Inc., Arlington, Va., private communication, May 17, 1994; Andrew Mollet, "Portugal's Electricity Industry Faces Big Shake Up," *Electricity International*, August 1993; Holmes, op. cit., note 22; Tim Woolf, "Retail Competition in the Electric Industry: Lessons from the United Kingdom," Tellus Institute, Boston, Mass., February 1994.

83. James B. Rouse, "Beyond Retail Wheeling: Competitive Sourcing of Retail Electric Power," *The Electricity Journal*, April 1994.

84. For a history of the Great Britain experience, see Holmes, op. cit., note 22, and Woolf, op. cit., note 82.

85. Woolf, op. cit., note 82; National Consumer Council, "Electricity Distribution: Price Control, Reliability and Customer Services," London, February 1994.

86. Norway from Armond Cohen, senior attorney, Conservation Law Foundation, "Retail Wheeling and Rhode Island's Energy Future: Issues, Problems, and Lessons from Europe," presented to the Retail Wheeling Subcommittee of the Rhode Island Energy Coordinating Council, July 22, 1993; State of California, Public Utilities Commission (CPUC), "Order Instituting Rulemaking on the Commission's Proposed Policies Governing Restructuring California's Electric Services Industry and Reforming Regulation," San Francisco, Calif., April 20, 1994; Seth Mydans, "California Nears Competition Among Electricity Providers," *New York Times*, April 21, 1994; Ralph Cavanagh, "California PUC Proposal to Restructure California's Electric Power Industry," memo, NRDC, San Francisco, Calif., April 23, 1994; Ralph Cavanagh, NRDC, San Francisco, Calif., private communication, May 18, 1994.

87. Silverman, op. cit., note 26; Armond Cohen and Steven Kihm, "The Political Economy of Retail Wheeling, or How to Not Re-Fight the Last War," *The Electricity Journal*, April 1994.

88. Cohen and Kihm, op. cit., note 87.

89. Employment from Cohen, op. cit., note 86; CPUC, op. cit., note 86.

90. Amory B. Lovins, Rocky Mountain Institute, Snowmass, Colo., letter to Daniel Fessler, President, CPUC, May 22, 1994.

91. Figure 7 is based on end-of-month closes from *S&P Security Price Index Record: Statistical Service* (New York: McGraw-Hill, 1992), from *S&P Statistical Service: Current Statistics* (New York: McGraw-Hill, 1994), from *Barron's National Business* *and Financial Weekly*, May 6, 1994, and from *Wall Street Journal*, June 1, 1994; Moody's Investors Service, "High-Cost Producers at Risk as California Opens its Electricity Market," New York, N.Y., April 1994; California utility stock prices from IDD Information Services, *Tradeline* (electronic database), Waltham, Mass., May 24, 1994; market capitalization loss is from Gloria Quinn, Edison Electric Institute, private communication, May 20, 1994.

92. Agis Salpukas, "Faster Cost Pass-Through for California Utility," *New York Times*, May 26, 1994; "California Agency Recommends Shutdown of San Onofre-2 and -3," *The Nuclear Monitor*, May 9, 1994; Shearson Lehman Brothers, "Should Investors Be Concerned About Rising Nuclear Plant Decommissioning Costs?" New York, N.Y., January 6, 1993.

93. Cavanagh, op. cit., note 61.

94. CPUC, op. cit., note 86.

95. Cavanagh, op. cit., note 61; Cohen and Kihm, op. cit., note 87; Bronwen Maddox, "Row over Saving Energy Intensifies," *Financial Times*, April 28, 1994; Hugh Outhred, University of New South Wales, Australia, "Achieving Least Cost Outcomes in the Emerging Competitive Electricity Industry," to be presented at the Second National Demand Management and Energy Efficiency Conference, Canberra, July 11-13, 1994.

96. California from Eric Miller, Kenetech Windpower, San Francisco, Calif., private communication, May 24, 1994; Niagara Mohawk is from Terry Black, Pace University Center for Environmental Legal Studies, White Plains, N.Y., private communication, May 27, 1994; renewable electricity demonstration proposals in New England also have encountered skeptical regulators, who are anticipating the onset of retail wheeling in their jurisdictions, according to Jeffrey Tranen, System Vice President, New England Electric System, Westborough, Mass., private communication, May 19, 1994.

97. Among the many proposals for electric utility restructuring made in recent years, the ones closest to ours are found in Outhred, op. cit., note 95, and in Stephen Wiel, "Achieving the Outcomes of Integrated Resource Planning within the Restructured Australian Electricity Industry," Lawrence Berkeley Laboratory, Washington, D.C., November 22, 1993.

98. U.S. Congress, op. cit., note 20.

99. See, for example, "NU, CMP Settle Differences in Wholesale Power Dispute," *The Energy Daily*, May 18, 1994, Tux Turkel, "Town Shops for Best Deal for Electricity," *Maine Sunday Telegram*, July 25, 1993, and Caleb Solomon, "As Competition Roils Electric Utilities, They Look to New Mexico," *Wall Street Journal*, May 9, 1994.

100. Sara Patton, Northwest Conservation Act Coalition, Seattle, Wa., private communication, May 26, 1994; Jeff Shields, Emerald People's Utility District, Eugene, Oreg., private communication, May 27, 1994.

101. Sacramento Municipal Utility District (SMUD), *1992 Annual Report: Working for Sacramento* (Sacramento, Calif.: undated).

102. Ibid; transcript from "Living on Earth," National Public Radio, August 27, 1993; SMUD, "Official Statement Relating to Sacramento Municipal Utility District $498,410,000 Electric Revenue Refunding Bonds, 1993 Series D $75,000,000 Electric Revenue Bonds, 1993 Series E," Sacramento, Calif., April 15, 1993.

103. Moskovitz, op. cit., note 58.

104. Ibid.

105. Cohen, op. cit., note 86; Don Bain, "New Northwest Resources: Gas, or Conservation and Renewables?" Oregon Department of Energy, Salem, Oreg., December 23, 1992; Flavin and Lenssen, op. cit., note 39; New England Electric System, *NEESPLAN 4* (Westborough, Mass.: 1993).

106. Pace University Center for Environmental Legal Studies, *Environmental Costs of Electricity* (New York: Oceana Publications, 1990); Hamrin and Rader, op. cit., note 36; Great Britain from Jane Massy, "Renewables Target Increased," *Windpower Monthly*, May 1993; David H. Moskovitz, "Green Pricing: Experience and Lessons Learned," The Regulatory Assistance Project, Gardiner, Maine, undated.

107. Levine et al., op. cit., note 64; OTA, *Fueling Development: Energy Technologies for Developing Countries* (Washington, D.C.: GPO, 1992); Nicholas Lenssen, *Empowering Development: The New Energy Equation*, Worldwatch Paper 111 (Washington, D.C.: Worldwatch Institute, November 1992).

THE WORLDWATCH PAPER SERIES

No. of
Copies

_____ 57. **Nuclear Power: The Market Test** by Christopher Flavin.
_____ 58. **Air Pollution, Acid Rain, and the Future of Forests** by Sandra Postel.
_____ 60. **Soil Erosion: Quiet Crisis in the World Economy** by Lester R. Brown and Edward C. Wolf.
_____ 61. **Electricity's Future: The Shift to Efficiency and Small-Scale Power** by Christopher Flavin.
_____ 62. **Water: Rethinking Management in an Age of Scarcity** by Sandra Postel.
_____ 63. **Energy Productivity: Key to Environmental Protection and Economic Progress** by William U. Chandler.
_____ 65. **Reversing Africa's Decline** by Lester R. Brown and Edward C. Wolf.
_____ 66. **World Oil: Coping With the Dangers of Success** by Christopher Flavin.
_____ 67. **Conserving Water: The Untapped Alternative** by Sandra Postel.
_____ 68. **Banishing Tobacco** by William U. Chandler.
_____ 69. **Decommissioning: Nuclear Power's Missing Link** by Cynthia Pollock.
_____ 70. **Electricity For A Developing World: New Directions** by Christopher Flavin.
_____ 71. **Altering the Earth's Chemistry: Assessing the Risks** by Sandra Postel.
_____ 73. **Beyond the Green Revolution: New Approaches for Third World Agriculture** by Edward C. Wolf.
_____ 74. **Our Demographically Divided World** by Lester R. Brown and Jodi L. Jacobson.
_____ 75. **Reassessing Nuclear Power: The Fallout From Chernobyl** by Christopher Flavin.
_____ 76. **Mining Urban Wastes: The Potential for Recycling** by Cynthia Pollock.
_____ 77. **The Future of Urbanization: Facing the Ecological and Economic Constraints** by Lester R. Brown and Jodi L. Jacobson.
_____ 78. **On the Brink of Extinction: Conserving The Diversity of Life** by Edward C. Wolf.
_____ 79. **Defusing the Toxics Threat: Controlling Pesticides and Industrial Waste** by Sandra Postel.
_____ 80. **Planning the Global Family** by Jodi L. Jacobson.
_____ 81. **Renewable Energy: Today's Contribution, Tomorrow's Promise** by Cynthia Pollock Shea.
_____ 82. **Building on Success: The Age of Energy Efficiency** by Christopher Flavin and Alan B. Durning.
_____ 83. **Reforesting the Earth** by Sandra Postel and Lori Heise.
_____ 84. **Rethinking the Role of the Automobile** by Michael Renner.
_____ 85. **The Changing World Food Prospect: The Nineties and Beyond** by Lester R. Brown.
_____ 86. **Environmental Refugees: A Yardstick of Habitability** by Jodi L. Jacobson.
_____ 87. **Protecting Life on Earth: Steps to Save the Ozone Layer** by Cynthia Pollock Shea.
_____ 88. **Action at the Grassroots: Fighting Poverty and Environmental Decline** by Alan B. Durning.
_____ 89. **National Security: The Economic and Environmental Dimensions** by Michael Renner.
_____ 90. **The Bicycle: Vehicle for a Small Planet** by Marcia D. Lowe.
_____ 91. **Slowing Global Warming: A Worldwide Strategy** by Christopher Flavin
_____ 92. **Poverty and the Environment: Reversing the Downward Spiral** by Alan B. Durning.
_____ 93. **Water for Agriculture: Facing the Limits** by Sandra Postel.
_____ 94. **Clearing the Air: A Global Agenda** by Hilary F. French.
_____ 95. **Apartheid's Environmental Toll** by Alan B. Durning.
_____ 96. **Swords Into Plowshares: Converting to a Peace Economy** by Michael Renner.
_____ 97. **The Global Politics of Abortion** by Jodi L. Jacobson.
_____ 98. **Alternatives to the Automobile: Transport for Livable Cities** by Marcia D. Lowe.
_____ 99. **Green Revolutions: Environmental Reconstruction in Eastern Europe and the Soviet Union** by Hilary F. French.

_____100. **Beyond the Petroleum Age: Designing a Solar Economy** by Christopher Flavin and Nicholas Lenssen.
_____101. **Discarding the Throwaway Society** by John E. Young.
_____102. **Women's Reproductive Health: The Silent Emergency** by Jodi L. Jacobson.
_____103. **Taking Stock: Animal Farming and the Environment** by Alan B. Durning and Holly B. Brough.
_____104. **Jobs in a Sustainable Economy** by Michael Renner.
_____105. **Shaping Cities: The Environmental and Human Dimensions** by Marcia D. Lowe.
_____106. **Nuclear Waste: The Problem That Won't Go Away** by Nicholas Lenssen.
_____107. **After the Earth Summit: The Future of Environmental Governance** by Hilary F. French.
_____108. **Life Support: Conserving Biological Diversity** by John C. Ryan.
_____109. **Mining the Earth** by John E. Young.
_____110. **Gender Bias: Roadblock to Sustainable Development** by Jodi L. Jacobson.
_____111. **Empowering Development: The New Energy Equation** by Nicholas Lenssen.
_____112. **Guardians of the Land: Indigenous Peoples and the Health of the Earth** by Alan Thein Durning.
_____113. **Costly Tradeoffs: Reconciling Trade and the Environment** by Hilary F. French.
_____114. **Critical Juncture: The Future of Peacekeeping** by Michael Renner.
_____115. **Global Network: Computers in a Sustainable Society** by John E. Young.
_____116. **Abandoned Seas: Reversing the Decline of the Oceans** by Peter Weber.
_____117. **Saving the Forests: What Will It Take?** by Alan Thein Durning.
_____118. **Back on Track: The Global Rail Revival** by Marcia D. Lowe.
_____119. **Powering the Future: Blueprint for a Sustainable Electricity Industry** by Christopher Flavin and Nicholas Lenssen.

_____ **Total Copies**

☐ **Single Copy: $5.00**
☐ **Bulk Copies (any combination of titles)**
☐ 2–5: $4.00 ea. ☐ 6–20: $3.00 ea. ☐ 21 or more: $2.00 ea.

☐ **Membership in the Worldwatch Library: $30.00 (international airmail $45.00)**
The paperback edition of our 250-page "annual physical of the planet," *State of the World 1994*, plus all Worldwatch Papers released during the calendar year.

☐ **Subscription to *World Watch* Magazine: $20.00 (international airmail $35.00)**
Stay abreast of global environmental trends and issues with our award-winning, eminently readable bimonthly magazine.

Please include $3 postage and handling per order.

Make check payable to Worldwatch Institute
1776 Massachusetts Avenue, N.W., Washington, D.C. 20036-1904 USA

Enclosed is my check for U.S. $_______

____
name **daytime phone #**

____
address

____
city **state** **zip/country**

WWP